Microbial Photosynthesis

Rachapudi V. Sreeharsha • S. Venkata Mohan

Microbial Photosynthesis

From Basic Biology to Artificial Cell Factories and Industrial Applications

Springer

Rachapudi V. Sreeharsha
Department of Life Sciences
CSJM Kanpur University
Kanpur, India

S. Venkata Mohan
Department of Energy and Environmental Engineering
CSIR-Indian Institute of Chemical Technology (CSIR-IICT)
Hyderabad, India

ISBN 978-981-97-1252-6 ISBN 978-981-97-1253-3 (eBook)
https://doi.org/10.1007/978-981-97-1253-3

This Springer imprint is published by the registered company Springer Nature Singapore Pte Ltd.
The registered company address is: 152 Beach Road, #21-01/04 Gateway East, Singapore 189721, Singapore

Dedicated To
The Family

Preface

In the intricate annals of biological processes, photosynthesis is one of the fascinating and scientifically captivating phenomena. *Microbial Photosynthesis: From Basic Biology to Artificial Cell Factories and Industrial Applications*, delve into the very essence of this remarkable phenomenon. This meticulously crafted book unfurls in three distinct parts, each encapsulating a facet of microbial photosynthesis. The chapters in Part I explore deep into the evolution, classification, and physiology of photosynthetic microbes. From the structural intricacies of microbial photosystem complexes to the elegant mechanisms of photosynthetic CO_2 fixation in unicellular organisms, these chapters serve as the foundation for applied photosynthesis. Part II ascends to the next level of comprehension, wherein the chapters investigate the practical applications of microbial photosynthesis. Techniques in photosynthetic physiology and innovative cultivation strategies are unveiled, laying the basis for sustainable resource utilization. The chapters dedicated to wastewater treatment and nutrient recovery, through cyanobacteria and microalgae, illuminate how these microorganisms can spearhead the charge towards a more sustainable future. Part III deals with the cutting-edge horizons of microbial photosynthesis. Genetic and metabolic engineering, as presented in Chap. 8, are the tools of the future to enhance the efficiency and potential of this biological process. The concept of industrial symbiosis, as elucidated in Chap. 9, sheds light on the circular economy forged by photosynthetic microbial biorefineries. And Chap. 10 ventures into the frontier of artificial photosynthesis—a complex interplay between photon energy and bioreactors that holds the promise of revolutionizing global energy landscape.

Throughout the book, the readers will encounter the findings and opinions of distinguished experts whose research contributions have provided valuable insights on photosynthesis and reflect a vision of the future. Our perspectives on the research topics of this book are helpful to transform industries and catalyze environmental sustainability. The contents of this book are highly useful to graduate students and scholars working in the areas of improving photosynthetic efficiency of microbes.

Kanpur, Uttar Pradesh, India — Rachapudi V. Sreeharsha
Hyderabad, Telangana, India — S. Venkata Mohan

Acknowledgments

This work is a result of several scientific contributions made by various research groups around the world. The authors thank all those who contributed significantly to the knowledge of microbial photosynthesis and allied research areas. RVS acknowledges guidance from his mentor Prof. Attipalli R Reddy, Emeritus Professor, University of Hyderabad.

Sincere thanks to Dr. Shalini Mudalkar, Assistant Professor, Tree Breeding and Improvement, Forest College and Research Institute, Hyderabad for her critical evaluation, invaluable suggestions, and diligent help throughout the work. Thanks to Dr. Ravi Kiran Boda, Post Doctoral Fellow, CSIR-IICT, Hyderabad; Dr. Aravind Kumar Konda, Scientist, ICAR-Indian Institute of Pulses Research (ICAR-IIPR), Kanpur; Dr. VNS Deviram Garlapati, Project Scientist, MoES, New Delhi; Dr. Divya K Unnikrishnan, University of Hyderabad, for their inputs and for reviewing the contents. The help of our students Ms. Sai Tejaswini, Mr. Akhil Singh, and Ms. Payal in the preparation of this manuscript is highly acknowledged.

We thank funding from the INSPIRE Faculty grant (DST/INSPIRE/04/2018/000214) from DST—SERB, New Delhi (RVS) and the Tata Innovation Fellowship grant (No. BT/HRD/35/01/02/2018) from DBT, New Delhi (SVM). We thank CSIR-IICT, Hyderabad and CSJMU, Kanpur for supporting this work.

Contents

About the Authors

Rachapudi V. Sreeharsha is currently working as Assistant Professor at the Department of Life Sciences, CSJM Kanpur University, Kanpur. He worked as DST-INSPIRE Faculty at CSIR-Indian Institute of Chemical Technology (CSIR-IICT). He completed his B.Sc. (Biotechnology) and M.Sc. (Molecular Genetics) from Andhra University, India and obtained his Ph.D. in Plant Sciences from the University of Hyderabad, India. His research interests include photosynthesis, climate change mitigation, and biofuels. His research contributed significantly to understand the molecular responses of legume crops to elevated atmospheric CO_2 concentrations. Currently, he is working on improving second and third generation biodiesel feedstock for their growth, photosynthesis, and productivity through metabolic engineering approaches. He received Best Thesis Award for his doctoral work from the University of Hyderabad and received Scientist of the Year Award 2022 from the National Environmental Science Academy (NESA), New Delhi. He is "Associate Fellow" of Telangana Academy of Sciences, Hyderabad.

S. Venkata Mohan is currently the Chief Scientist at the CSIR-Indian Institute of Chemical Technology, Hyderabad, India. He was a Visiting Professor at Kyoto University in 2005 and an Alexander von Humboldt (AvH) Fellow at the Technical University of Munich, Germany, from 2001 to 2002. Dr. Mohan's research addresses human-induced environmental changes within a sustainability framework at the intersection of environment and bioengineering. His current focus is on developing sustainable processes through a nexus approach for the production of low-carbon energy, chemicals, and materials with a regenerative and resilient approach. His research interests also include advanced waste remediation, biorefinery, environmental surveillance, and Circular Bioeconomy. Dr. Mohan has published 450 research articles, 60 book chapters, edited 5 books, holds 14 patents, and has supervised 38 Ph.D. students. His work has received over 33,000 citations with an h-index of 100 on Google Scholar. He is the recipient of the prestigious Shanti Swarup Bhatnagar (SSB) Prize from the Government of India and has received several other awards and honors, including the INAE-SERB Abdul Kalam Technology Innovation National Fellowship, DBT-Tata Innovation Fellowship (2018), the Prosper.net-Scopus Young Researcher Award by the United Nations University, etc. Dr. Mohan is an elected Fellow of the National Academy of Engineering, the Telangana and Andhra Pradesh Academy of Sciences, and serves on the editorial boards of various journals such as Transactions of the Indian National Academy of Engineering (Springer), Journal of Energy (Wiley), Bioresource Technology, Environmental Technology, etc.

Abbreviations

3HP-4HB	3-hydroxypropionate-4-hydroxybutyrate
ABS/RC	Absorption Flux per Reaction Center
ACCase	Acetyl-CoA carboxylase
AEM	Anion exchange membrane
ALRs	Airlift reactors
ARDS	Acute respiratory distress syndrome
ARE	Antioxidant responsive element
ASFV	African swine fever virus
ATP	Adenosine triphosphate
BBES	Balanced Bioenergy System
BBNC	Bioprocess and Biorefinery Nexus Consortium
BESs	Bioelectrochemical systems
BET	Bioelectrochemical treatment
BOD	Biochemical oxygen demand
C_{10}	Geranyl diphosphate
C_{15}	Farnesyl diphosphate
C_{20}	Geranylgeranyl diphosphate
CAM	Crassulacean acid metabolism
Car	Carotenoids
CBCRs	Cyanobacteriochromes
CCA	Chromatic acclimation
CCM	CO_2 concentrating mechanism
CEF	Cyclic electron flow
CEM	Cation exchange membrane
Chl	Chlorophyll
Chlide	Chlorophyllide
COD	Chemical oxygen demand
Copro	Coproporphyrinogen
CpSRP	Chloroplast signal recognition particle
CRISPR	Clustered regularly interspaced short palindromic repeats
Cyt	Cytochrome

DAO	Direct anodic oxidation
DCMU	(3-(3,4-dichlorophenyl)-1,1-dimethylurea)
DHA	Docosahexaenoic acid
DI0/RC	Dissipated Energy Flux per Reaction Center
DMAPP	Dimethylallyl pyrophosphate
DOE	Department of Energy
DPA	Docosapentaenoic acid
DSA	Dimensionally stable anodes
EAB	Electrochemically active bacteria
EFCs	Enzymatic fuel cells
EMCV	Encephalomyocarditis RNA virus
EPA	Eicosapentaenoic acid
EPS	Exopolysaccharides
ETC	Electron transfer chain
ETC/RC	Electron Transport Flux per Reaction Center
F_0	Initial fluorescence
F_0/F_m	Fluorescence parameter
FACE	Free air CO_2 enrichment
FAD	Flavin adenine dinucleotide
FADGDH	Flavin adenine dinucleotide-dependent glucose dehydrogenases
FaRLip	Far red-light photoacclimation
FBPase	Fructose 1,6-bisphosphatase
Fd	Ferredoxin
FDPs	Flavodiiron proteins
F_J	Fluorescence value
FLIM	Fluorescence lifetime imaging microscopy
Fm	Measure maximum fluorescence
FNR	Ferredoxin-$NADH^+$ reductase
FP	Peak fluorescence level
Ft	Measure steady-state fluorescence yield during light exposure
Fv	Fluorescence
Fx, F_A, F_B	Iron sulfur clusters
G_3P	Glyceraldehyde-3-phosphate
GAG	Glycosaminoglycans
GAPDH	Glyceraldehyde-3-phosphate dehydrogenase
GLA	γ-linolenic acid
Glu-R	Glutamyl-tRNA reductase
GOGAT	Glutamate synthase
GS	Glutamine synthetase
GSA	Glutamic acid 1-semialdehyde
GST-α1	Glutathione-S-transferase-α1
HBV	Hepatitis B virus
HHV-3	Human α herpesvirus 3
H_2S	Hydrogen sulfide

HMF	Hydroxymethylfurfural
HO-1	Heme oxygenase
HRAP	High-rate algal pond
HRT	Hydraulic retention time
HSV	Herpes simplex virus
HTA	Hexadecatrienoic acid
IPP	Isopentenyl diphosphate
IRGA	Infrared gas analyzer
LER	Light-enhanced respiration
LET/CET	Linear/Cyclic electron transfer
LHC	Light-harvesting complex
LHCR	LHC receptors
LPS	Lipopolysaccharide
MAAs	Mycosporine-like amino acids
MAP	Mitogen-activated protein
MDCs	Microbial desalination cells
MECs	Microbial electrolysis cells
MEP	Methyl erythritol 4-phosphate
MES	Microbial electrosynthesis
METs	Manufacturing of bio-based goods
MFCs	Microbial fuel cells
MMe	Monomethyl ester
MRSA	Methicillin-resistant *Staphylococcus aureus*
MSCs	Microbial solar cells
MUFA	Mono-unsaturated fatty acids
MuLV	Murine leukemia virus
MVA	Mevalonate
NADH/NADPH	Nicotinamide adenine dinucleotide/phosphate hydrogen
NF-KB	Nuclear transcription factor-KB
NIR	Near-infrared fluorescence
NPQ	Non-photochemical chlorophyll fluorescence quenching
NQO-1	Quinone oxidoreductase
NSAIDs	Non-steroidal anti-inflammatory drugs
OEC	Oxygen-evolving complex
OMBRs	Osmotic membrane bioreactors
PA	Palmitoleic acid
PAM	Pulse amplitude modulation
PBG	Porphobilinogen
PBS	Phycobilisomes
PC	Plastocyanin
Pchilde	Protochlorophyllide
PHA	Polyhydroxyalkanoate
PHB	Polyhydroxy butyrate
Pi	Phosphate
PI	Performance index

PI_{ABS}	Performance index
PMBC	Peripheral blood mononuclear cells
PMFCs	Plant microbial fuel cells
PPFD	Photosynthetic photon flux density
PQQ-ADH	Pyrroquinoline quinone-dependent alcohol dehydrogenase
PRK	Phosphoribulokinase
Proto	Protoporphyrin
Protogen	Protoporphyrinogen
PS	Polysaccharides
PS I	Photosystem I
PS II	Photosystem II
PsaA	PSI protein A
PsaB	PSI protein B
PTOX	Plastoquinone (plastid) terminal oxidase
PUFA	Poly-unsaturated fatty acids
PVC	Polyvinyl chloride
Q_A	Quinone acceptor
Q_B	Plastoquinone B
QBH_2	Reduced hydroquinone
ROS	Reactive oxygen species
RPP	Reductive pentose phosphate cycle
RSV-A/B	Respiratory syncytial virus
RuBISCO	Ribulose-1,5-bisphosphate carboxylase/oxygenase
RuBP	Ribulose-1,5 bisphosphate
SBPase	Sedoheptulose-1,7-bisphosphate
SCP	Single cell protein
TAG	Triacylglycerols
TALENs	Transcription activator like endonucleases
TR_0/RC	Trapped Energy Flux per Reaction Center
UM	Urban metabolism
Uro	Uroporphyrinogen
VACV	Vaccinia virus
VEGF	Vascular endothelial growth factor
VHSV	Hemorrhagic septicemia virus
VSV	Vesicular stomatitis virus
ZFN	Zinc-finger nuclease
ΔpH	Trans-thylakoid proton gradient
$\varphi D0$	Quantum Yield of Non-Radiative Energy Dissipation
$\varphi E0$	Quantum Yield of Electron Transport
$\varphi P0$	Quantum Yield of Primary Photochemistry

Part I
Photochemistry of Microbial Photosynthesis

Chapter 1
Photosynthetic Microbes: Evolution, Classification, and Structural Physiology

Abstract Photosynthesis converts carbon dioxide and water into organic molecules and oxygen by using light energy and provides the basic building blocks for all the life on Earth. It takes place in chloroplasts of green plants, cyanobacteria, and algae. The chlorophyll pigment in chloroplasts absorbs light energy, which is then used to split water molecules into oxygen and hydrogen ions. The hydrogen ions are used to generate ATP, which provides energy for the synthesis of glucose from carbon dioxide through Calvin cycle. Other than Calvin cycle, there exists different routes including reverse Krebs cycle, and 3-hydroxypropionate-4-hydroxybutyrate cycle through which carbon fixation occurs in certain bacteria and archaea. These cycles have potential applications in biotechnology and bioengineering to produce biofuels and other bioproducts. Cyanobacteria and microalgae are unicellular photosynthetic microbes belonging to diverse taxonomical groups. Nevertheless, there are certain similarities between the two groups, for instance in pigment biosynthesis and functioning of photosystem complexes. Studying the photosynthetic physiology in these microbes provides crucial insights into the evolution of oxygenic photosynthesis in higher plants and its practical implications. Their ability to grow in unfavorable conditions and to utilize waste resources made them valuable industrial cell factories. The chapter highlights the photosynthetic microbes and their cellular structure and their complexity for photosynthesis process.

Keywords Oxygenic photosynthesis · Anoxygenic photosynthesis · 3-hydroxypropionate-4-hydroxybutyrate · Reverse Krebs cycle · Calvin cycle · Cyanobacteria

1.1 Introduction

Photosynthesis has played a pivotal role in altering the Earth's atmosphere to support the life forms. The evolution of photosynthesis can be dated back to billions of years, and it is considered the most dynamic and transformative process. When the Earth formed around 4.6 billion years ago, it was primarily composed of volcanic

R. V. Sreeharsha, S. Venkata Mohan, *Microbial Photosynthesis*,
https://doi.org/10.1007/978-981-97-1253-3_1

gases, such as carbon dioxide (CO_2), nitrogen (N_2), methane (CH_4), and water vapor with no free oxygen (O_2). Eventually, few early life forms practiced anoxygenic photosynthesis by using substances such as hydrogen sulfide, iron, or organic molecules as electron donors nevertheless without producing oxygen. One of the fundamental developments in the evolution of photosynthesis occurred around 3.5 to 2.5 billion years ago when cyanobacteria, also known as blue-green algae emerged. These prokaryotic microorganisms were among the first to perform oxygenic photosynthesis, a process that uses water and releases oxygen as a by-product. Initially, the oxygen produced during photosynthesis was absorbed by dissolved iron in the oceans and other chemical reactions, preventing it from significantly accumulating in the atmosphere. Eventually, the oxygen produced by cyanobacteria began to accumulate in the atmosphere, leading to the "Great Oxygenation Event" around 2.4 to 2.3 billion years ago. This event had a profound impact on life on Earth, leading to the extinction of many anaerobic organisms (organisms that cannot survive in the presence of oxygen) and paving the way for the rise of aerobic organisms (organisms that require oxygen to live). Also, this event marked a significant transition from a reducing atmosphere to an oxidizing one.

Around 1.5 billion years ago, a eukaryotic cell engulfed a cyanobacterium through endosymbiosis, which was a major event in the process of evolution. Over time, this cyanobacterium evolved into chloroplasts, the organelles responsible for photosynthesis in plants and algae, thus allowing eukaryotes to harness the power of photosynthesis. After the endosymbiotic event, photosynthetic organisms diversified, leading to a wide variety of groups that we see today. Each group developed unique adaptations and strategies for photosynthesis, enhancing the efficiency of the process. The Phanerozoic Era, which started around 541 million years ago and continues up until the present day, saw a further rise of oxygen levels in the atmosphere. As photosynthetic organisms continued to thrive, they absorbed carbon dioxide and released oxygen, significantly influencing the composition of the atmosphere, playing a critical role in the carbon cycle and providing food for countless species (Sreeharsha and Venkata Mohan 2021). Furthermore, the unicellular and multicellular photosynthetic microorganisms currently scattered in the terrestrial and aquatic biome along with green plants are involved in the sequestration of the persistently increasing CO_2 levels in the atmosphere thus mitigating the global warming (Kumar et al. 2017; Madhana Sekhar et al. 2014; Sreeharsha et al. 2019).

This chapter gives an overview of anoxygenic and oxygenic photosynthesis while discussing the structural features of the microbes that perform oxygenic photosynthesis which is the crux of the book. Also, the chapter discusses in detail about various C sequestration mechanisms prevailing in nature which is crucial for the photosynthesis process. The reader must understand the complexities of photosynthetic pigments in microbes to decipher the light capturing, photolysis of water, and C sequestration which are discussed in detail in the upcoming chapters.

1.2 Anoxygenic Photosynthesis

The photochemical reaction that is performed by certain groups of bacteria without producing oxygen and without the utilization of water molecule as an electron donor is considered as anoxygenic photosynthesis. The global impact of anoxygenic photosynthesis on gaseous cycles is considered to be negligible. The existence of anoxygenic photosynthetic bacteria was first identified in 1901 in *Rhodospirillum* species and further research has spread the phylogenetic extent with the identification of various other species. Compared to the oxygenic photosynthesis which is limited to only one phylum, i.e., *Cyanobacteria*, the other species in the bacterial domain are known to perform anoxygenic photosynthesis (Fig. 1.1). The species that perform anoxygenic photosynthesis were distinguished by their habitats and their photosynthetic apparatus. Specifically, the composition of photosynthetic pigments and light-harvesting complexes make them unique from other species. Bacteriochlorophylls which are similar to chlorophylls in structure but have a slightly different absorption spectrum play a major role in mediating anoxygenic photosynthesis by utilizing light that is not absorbed by chlorophyll-containing organisms. Different phyla of bacterial domain and the corresponding species that perform anoxygenic photosynthesis are described below.

1.2.1 Green Sulfur Bacteria

Green sulfur bacteria (Chlorobiaceae) are a group of photosynthetic bacteria that belong to the phylum Chlorobi. They are anaerobic, phototrophic bacteria that use bacteriochlorophylls and various carotenoids to carry out photosynthesis. Unlike plants and algae, these bacteria do not produce oxygen during photosynthesis and instead utilize sulfur compounds or organic compounds as electron donors. Green sulfur bacteria are prevalent in low-light conditions and may be found in a number of aquatic habitats, including freshwater and marine sediments. They are typically green in color due to the presence of *chlorosomes*, which are unique light-harvesting organelles that are capable of absorbing light from a broad range of wavelengths. Few species including *Chlorobium tepidum* are also capable of fixing atmospheric nitrogen, which allows them to grow in nitrogen-limited environments. Nitrogenase enzyme plays an important role in both mesophilic and thermophilic anoxygenic phototrophs. The genus *Prosthecochloris* belonging to phylum Chlorobi shows an exceptional ability to fix atmospheric N_2 and to perform anoxygenic photosynthesis and support the growth of nutrient-limited coral reef ecosystem (Yang et al. 2016). These microbes are crucial for the biogeochemical cycling of sulfur and carbon in the aquatic environments (Behera et al. 2014; Sirevåg and Ormerod 1970).

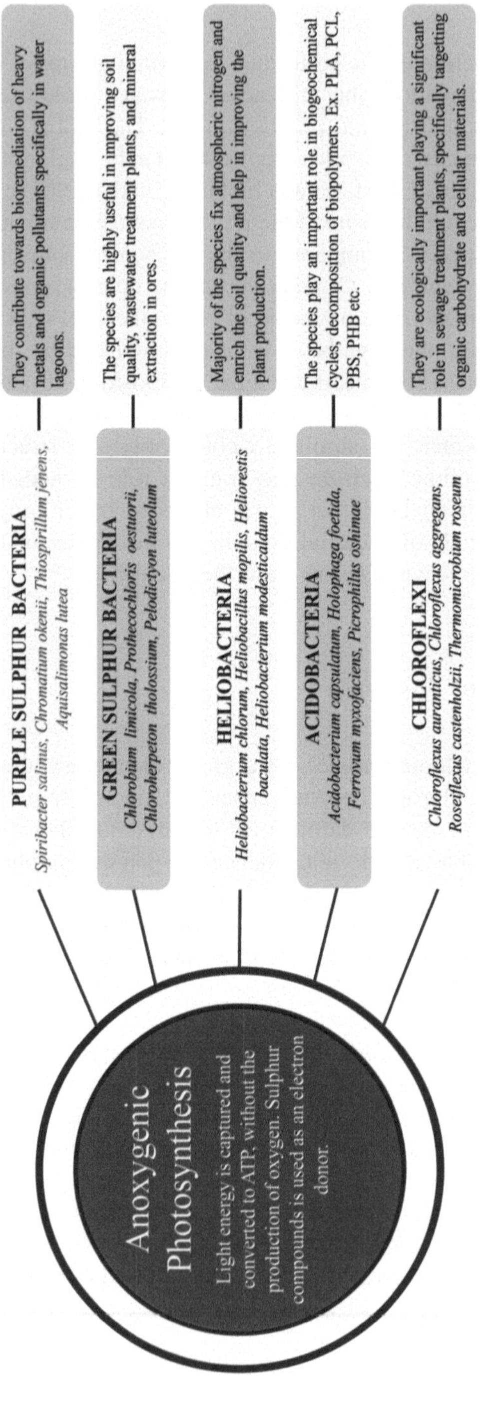

Fig. 1.1 Diverse groups of microbes that perform anoxygenic photosynthesis and their important species along with their industrial applications

1.2.2 Purple Sulfur Bacteria

Purple sulfur bacteria are anaerobic, photosynthetic bacteria that belong to the phylum Proteobacteria. Purple sulfur bacteria also possess *bacteriochlorophylls* which absorb light at longer wavelengths than chlorophyll, allowing them to carry out photosynthesis in low-light conditions. They also contain carotenoids, which protect them from damage caused by excess light. Purple sulfur bacteria are typically found in environments with high levels of organic matter, such as the bottom of stagnant water bodies, mud, or microbial mats. They use hydrogen sulfide (H_2S) as an electron donor during photosynthesis, producing sulfur granules as a by-product that gives them a distinctive smell. Purple sulfur bacteria are significant contributors to the productivity of aquatic habitats and play an essential role in the carbon and sulfur cycles of many ecosystems (Pfennig and Widdel 1982).

1.2.3 Heliobacteria

Heliobacteria are obligate anaerobes with a unique type of chlorophyll called *chlorophyll g*, which is structurally similar to bacteriochlorophylls but not found in any other known photosynthetic organism. They are typically found in soil environments and can perform photosynthesis under low-light conditions. They belong to the phylum Firmicutes and are typically found in soil and freshwater environments. Similar to cyanobacteria, algae, and higher plants, Heliobacteria possess photosystem I to generate ATP and reduce ferredoxin. Heliobacteria are unique in their ability to produce spores, which allows them to survive in harsh environmental conditions. They are also able to fix nitrogen, which makes them important contributors to nitrogen cycling in some ecosystems. Despite their ecological and evolutionary importance, Heliobacteria are not well studied compared to other groups of photosynthetic bacteria. However, they are of interest to researchers due to their potential for use in biotechnology applications, such as biofuel production and bioremediation (Heinnickel and Golbeck 2007).

1.2.4 Acidobacteria

Acidobacteria are considered to be one of the most abundant and diverse groups of bacteria in soil that can perform photosynthesis under low-light conditions. They are capable of performing anoxygenic photosynthesis using bacteriochlorophyll a or b, along with carotenoids. They are known for their metabolic diversity and the ability to thrive in diverse habitats, including acidic environments, soils, sediments, and aquatic environments. There are many different types of Acidobacteria, and they have a wide range of physiological and metabolic characteristics. Some are chemoorganotrophs, which use organic substances as both an energy source and a source of carbon, while others are photoheterotrophs, which use light as both energy and carbon source. Few Acidobacteria can fix nitrogen, while others can degrade complex organic compounds such as cellulose thus playing a crucial role in nutrient recycling. They are also important for plant growth, as they can form beneficial associations with plant roots and contribute to the availability of nutrients in the soil. Some of the species have also been studied for their potential use in bioremediation, as they can degrade toxic compounds and pollutants in soil and water (Dedysh and Damsté 2018) (Fig. 1.1).

1.2.5 Chloroflexi

Species of the Chloroflexi phylum are capable of anoxygenic photosynthesis using bacteriochlorophylls and carotenoids. They are typically found in hot springs and other environments with low oxygen and high sulfur concentrations. Chloroflexi is a phylum of bacteria that has a unique photosynthetic mechanism called chlorophototrophy, which allows them to use light as a source of energy to synthesize ATP. In addition, they make use of reduced sulfur compounds as electron donors, such as sulfide or thiosulfate, and they generate either sulfur or sulfite as end products. Chloroflexi are also known for their ability to grow under anaerobic conditions, as they can use alternative electron acceptors such as nitrate, ferric iron, or carbon dioxide in the absence of oxygen. This metabolic versatility makes them important players in carbon and nitrogen cycling in various environments, including soil, freshwater, and marine habitats. In addition to their unique photosynthetic and metabolic capabilities, Chloroflexi is also characterized by their filamentous or sheathed cell morphology, which allows them to form complex multicellular structures. Certain Chloroflexi species are especially known for their capacity to break down a wide variety of organic molecules, such as cellulose, lignin, and aromatic compounds, which makes them important players in the bioremediation of contaminated environments (Krzmarzick et al. 2012; Speirs et al. 2019) (Fig. 1.1). Thermophilic or mesophilic phototrophs which are considered as filamentous anoxygenic phototrophs because of their filamentous morphologies are also part of the Chloroflexi.

1.3 Oxygenic Photosynthesis

Oxygenic photosynthesis is the most abundant form of carbon fixation process that utilizes water molecule as electron donor and harness the sunlight to produce organic matter while simultaneously releasing the molecular oxygen. In the bacterial domain, Cyanobacteria largely perform oxygenic photosynthesis while microalgae are unicellular eukaryotic microorganisms that contribute significantly to the oxygenic photosynthesis especially in aquatic environments.

1.3.1 Cyanobacteria

Cyanobacteria are single-celled photosynthetic organisms that play a crucial role in the global carbon cycle and oxygen production. They display remarkable diversity, exhibiting unique cell morphologies, variable sizes, and different colors. They can also form complex filamentous groups called trichomes, which are covered in a slimy coating and resemble smooth threads (Singh et al. 2011). The size range of cyanobacteria varies widely, with unicellular Prochlorococcus being the smallest (0.7 μm cell diameter) and filamentous *Oscillatoria princeps* being the largest (55 μm cell width) (Ting 2014). The phylogenetic classification of Cyanobacteria identifies four orders including Chroococcales, Nostocales, Oscillatoriales, and Stigonematales (Table 1.1).

1.3.1.1 Structural Aspects and Cellular Mechanisms

Cyanobacteria exhibit a unique cell configuration compared to simpler bacteria and archaea (Waterbury 2006). The cell wall is an essential component of cyanobacteria that provides structural support and protection against environmental stresses. Their cell walls are made up of an exterior protective barrier of lipopolysaccharide (LPS) layer, as well as a thick underlying layer of peptidoglycan, that are separated by periplasm. Notably, the peptidoglycan layer that is present on cyanobacterial cells is approximately five times thicker than that of other Gram-negative bacteria, and the LPS is unique without any bound phosphate, heptoses, or keto-deoxyoctonate (Durai et al. 2015). In addition to these distinctive features, *Cyanobacteria* contain additional unique components in their outer membrane like carotenoids, hydroxylated fatty acids, and Toc75 homolog porins, which are important for protein translocation in photosynthetic organisms (Hoiczyk 2000; Nicolaisen et al. 2009). Surrounding the cyanobacteria cells there are sheaths, capsules, and slimes, which serve protective roles under extreme environmental conditions and aid in gliding movement (Sciuto and Moro 2015).

In filamentous cyanobacteria, the process of cellular differentiation results in the production of specialized cells such as heterocysts, diazocytes, akinetes, and hormogonia. Heterocysts are particularly important as they are responsible for nitrogen fixation in most species (Heimann and Cirés 2015; Oren 2014). They cannot fix carbon dioxide and must obtain carbon from adjacent vegetative cells. Heterocysts provide an appropriate micro-anaerobic condition and exhibit spatial separation for effective N_2 fixation. Additionally, heterocyst-forming cyanobacteria interact with eukaryotic organisms to develop symbiotic relationships, and approximately 8% of lichen species contain these kinds of associations (Cohen and Gurevitz 2006).

Table 1.1 Structural and physiological properties of selective important orders of cyanobacteria

Order	Important species	Habitat	Industrial applications
Chlorococcales	*Microcystis aeruginosa*	Fresh and brackish water	Production of butylated hydroxy toluene and bioactive peptides called Aerucyclamides
	Microcystis viridis	Freshwater with access phosphorus and nitrogen	Biofuel production, carbon capture and sequestration
Nostocales	*Nostoc commune*	Terrestrial and freshwater	Anti-inflammatory compounds and anti-oxidant activity
	Anabaena sp.	Freshwater	Bioremediation, biofertilizer and biofuel production
	Scytonema sp.	Freshwater	Cosmetics and personal care, pharmaceuticals, biofuel and biofertilizer
	Nostoc muscorum	Freshwater	Detoxify/accumulate heavy metals
Oscillatoriales	*Oscillatoria agardhii*	Freshwater	Nano-silicon products used on wheat crops reduces rust, powdery mildew, and spot diseases
	Lyngbya sp.	Brackish and marine water	Bioremediation, pigments, biofertilizers, biofuel, and pharmaceuticals
	Microcoleus sp.	Freshwater	Biofertilizer, biomineralization, biotechnology research, and biofuel production
Stigonematales	*Chlorogloeopsis fritschii*	Terrestrial	Biofuel and bioethanol production
	Fischerella muscicola	Freshwater	Production of high value compounds such as pharmaceutical or bioactive molecules

Diazocytes are distinctive nitrogen-fixing cells found in the genus *Trichodesmium* belonging to the order Oscillatoriales of *Cyanobacteria*. Akinetes are brown, thick-walled cyanobacterial spores while hormogonia are reproductive trichomes derived from terminal points of cyanobacteria. Akinetes are larger cells that contain reserve products such as glycogen and cyanophycin. They are enveloped in a thick protective coat composed of multilayered structures, which contribute to their resistance to adverse conditions. The formation of spore-like akinetes is a survival strategy employed by filamentous cyanobacteria under unfavorable environmental conditions.

Cyanobacteria have undergone chromatic adaptation with their chlorophylls and phycobilins in order to thrive in a varied illuminated environments (Komárek 2016; Shevela et al. 2013). Carotenoids, similar to higher plants, support light absorption and protect cyanobacterial cells from light damage (Koller 2015). Some cyanobacterial species contain alternative chlorophylls in their photosystems, expanding their photosynthetic capabilities (Trampe and Kühl 2016). Phycobilins, exclusive to cyanobacteria, red algae, and cryptomonads, also contribute to their diverse pigmentation (Chakdar and Pabbi 2016; Pinevich et al. 2012). The diverse pigmentation of cyanobacteria enables them to utilize the visible spectrum (400–700 nm) more efficiently than plants or algae, whose photosystems contain only chlorophylls and carotenoids.

Being a photosynthetic organism, Cyanobacteria share certain common features with higher plants and microalgae with respect to their chloroplast structure, biogenesis of thylakoids and protein loading onto the membrane. Thylakoid research in *Cyanobacteria* serve as model to understand the complex molecular mechanisms of the photosynthesis. Protein transport into the thylakoid membrane is a critical process in all the photosynthetic organisms. The SEC translocase systems are identified as major protein complexes that are responsible for protein import into chloroplasts (Frain et al. 2016). Cyanobacteria possess two distinct SEC translocase systems: The SEC1 system is found in the thylakoid membrane of cyanobacteria, whereas the SEC2 system is found in the inner envelope membrane. These systems have different modes of action and interact with various components to facilitate protein targeting, translocation, and integration. Cyanobacterial PetC3 Rieske proteins have been characterized structurally and functionally. They were initially identified as potential subunits of the cytochrome b6f complex but further research suggests that they have a distinct function related to cell envelope homeostasis rather than involvement in bioenergetic electron transfer. Chloroplast iron transport proteins play crucial roles in plant physiology by regulating metal homeostasis within chloroplasts. Chloroplasts require transition metals like iron for photosynthetic electron transport but must also prevent oxidative damage caused by reactive oxygen species generated during photosynthesis. Proteins involved in iron transport have been identified in the chloroplastic outer and inner envelope membranes, and their interactions with precursor proteins destined for transport into chloroplasts are being elucidated.

1.3.2 Microalgae

Microalgae are diversified unicellular photosynthetic organisms that evolved into well-defined taxonomical groups. The seven Phyla of kingdom Protista including Chlorophyta (Green Algae) (*Chlorella* and *Chlamydomonas*) Bacillariophyta (Diatoms), Haptophyta (Haptophytes) (*Emiliania huxleyi* and *Chrysochromulina* spp.), Dinoflagellata (Dinoflagellates) (*Alexandrium* and *Karenia*), Euglenophyta (Euglenoids) (*Euglena*), and Cryptophyta (Cryptomonads) (*Cryptomonas*). Important orders and their representative species are indicated in Fig. 1.2.

1.3.2.1 Structural and Cellular Aspects

The cell wall of microalgae is a vital structural component that surrounds the cell membrane, providing support, protection, and shape to the cell. The cell wall plays a significant role in the survival and function of microalgae in various

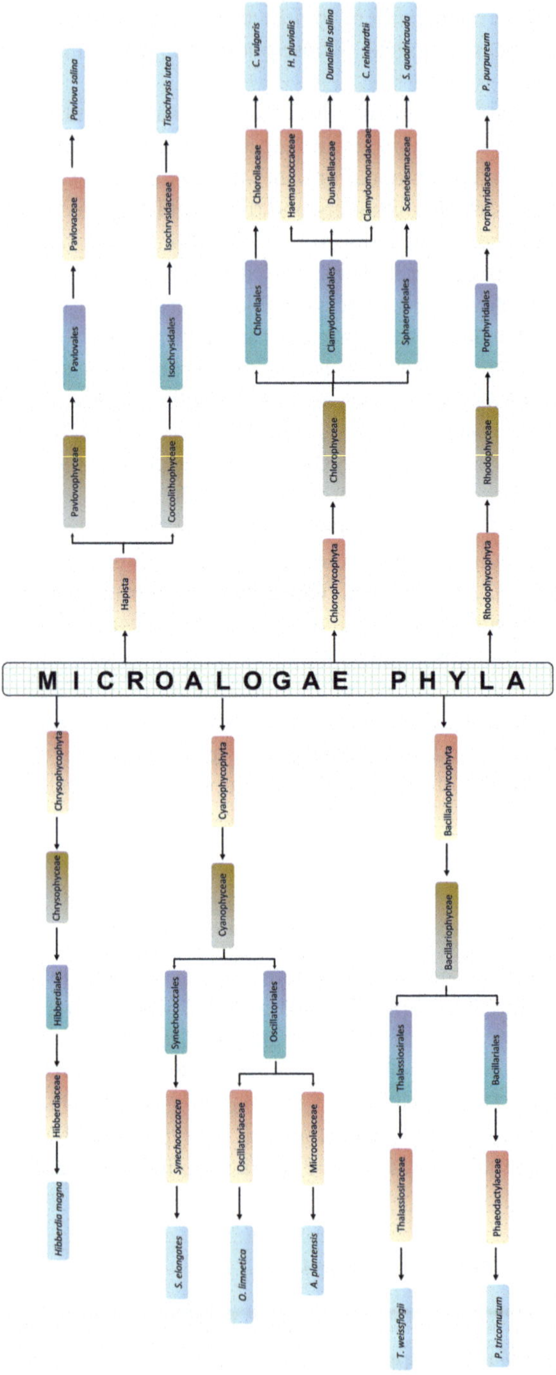

Fig. 1.2 Hierarchical classification of different phyla of microalgae

environmental conditions. The composition of the microalgae cell wall can vary widely among different species. Generally, the cell wall is primarily composed of complex polysaccharides, and proteins, and sometimes, in certain species, it may also contain minerals like silica. The main polysaccharides present in microalgae cell walls include cellulose, hemicellulose, and pectin (Ranadheer et al. 2019). Cellulose which is a linear polymer made up of glucose molecules provides strength and rigidity to the cell wall, making it resistant to mechanical stress and deformation. Hemicellulose is another group of polysaccharides with a more branched and flexible structure, adding flexibility and resilience to the cell wall. Pectin acts as a cementing material, binding cellulose fibers together and helping to form a gel-like matrix in the cell wall. It plays a role in cell adhesion and cell-to-cell communication. Some microalgae, particularly diatoms, have cell walls that contain silica. In diatoms, the cell wall consists of two overlapping halves, forming a shell-like structure called a frustule. Silica provides remarkable rigidity and protection to diatoms, making them abundant and diverse in various aquatic environments. Along with carbohydrates, cell wall proteins, glycoproteins, and glycolipids altogether regulate major events including cell signaling, defense against pathogens, cell recognition, adhesion, and communication. The microalgae cell wall is porous to some extent, allowing the passage of certain molecules into and out of the cell playing a role in nutrient uptake and waste removal.

The photosynthetic microbes belonging to diverse taxonomical groups contribute invaluably to photosynthetic research. The structural simplicity of these unicellular organisms enables research to understand the complex structural and functional aspects of photosynthesis and its applications. In addition, *Cyanobacteria* and microalgae are being considered as a promising source for alternative energy and other industrial applications thus meeting the increasing demand for renewable energy and the global fuel crisis (Sreeharsha et al. 2023). These photosynthetic microbes serve as a promising industrial feedstock with several advantages, such as minimal usage of arable land and a positive impact on mitigating elevated atmospheric CO_2 levels.

1.4 Carbon Fixation and Sources

Carbon fixation is the process by which inorganic carbon (usually carbon dioxide) is converted into organic compounds, such as sugars and other biomolecules. Popular carbon fixation pathways exist in nature are given in Fig. 1.3.

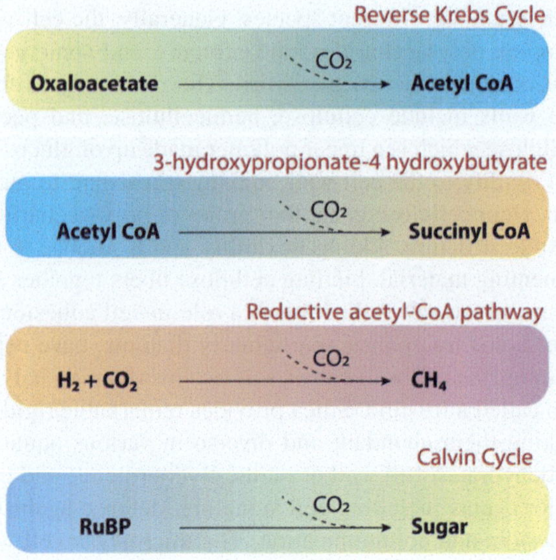

Fig. 1.3 Overview of different carbon fixation pathways exist in nature

1.4.1 Reverse Krebs Cycle

This pathway is found in autotrophic bacteria and archaea that live in extreme environments, such as hydrothermal vents. It uses carbon dioxide and other inorganic molecules to produce organic compounds. In the traditional Krebs cycle, CO_2 is released through decarboxylation reactions as acetyl-CoA is oxidized. In the reverse Krebs cycle, these reactions are reversed, and CO_2 is reduced and incorporated into an organic molecule, usually pyruvate or other intermediates, resulting in the formation of a new compound, oxaloacetate. The intermediates produced in the reverse Krebs cycle can be further processed, leading to the generation of acetyl-CoA, a central metabolite involved in various biosynthetic pathways. The reverse Krebs cycle allows certain microorganisms to fix carbon dioxide in environments where other carbon sources are limited or absent. It enables them to synthesize complex organic compounds from simple inorganic carbon sources, such as CO_2 (Buchanan and Arnon 1990; Muchowska et al. 2017).

1.4.2 3-Hydroxypropionate-4-Hydroxybutyrate Cycle

The 3-hydroxypropionate-4-hydroxybutyrate (3HP-4HB) cycle is a crucial metabolic pathway found in certain bacterial species living in extreme environments like hot springs, geysers, and deep-sea hydrothermal vents. This pathway plays a crucial role in carbon dioxide (CO_2) fixation, enabling the microorganisms to produce organic molecules for growth and cellular processes. The 3HP-4HB cycle is an important component in global carbon cycling, and its key genes have been found in various species, including *Metallosphaera*, *Sulfolobus*, *Archaeoglobus*, and *Cenarchaeum*. These organisms exhibit diverse energy metabolisms and habitats, ranging from volcanic areas to anaerobic sulfate-reducing environments. The 3HP-4HB cycle demonstrates relative insensitivity to oxygen, avoidance of side reactions, direct utilization of bicarbonate, and thermal stability, making it advantageous for autotrophic carbon assimilation in their habitats. The 3HP-4HB cycle was originally discovered in the phototrophic bacterium *Chloroflexus aurantiacus*, its subsequent exploration in Archaea revealed variations in certain key enzymes. In most Archaea, the acetyl-CoA/propionyl-CoA carboxylase enzyme seems to act on both acetyl-CoA and propionyl-CoA, indicating its promiscuous nature. The discovery of this cycle in an autotrophic archaeal species, *Metallosphaera sedula*, revealed the carboxylation of acetyl-coenzyme A (acetyl-CoA) to malonyl-CoA as a key step (Pearson et al. 2019). This process reductively converts one acetyl-CoA and two bicarbonate molecules via 3-hydroxypropionate to succinyl-CoA. The succinyl-CoA is further reduced to 4-hydroxybutyrate and converted into two molecules of acetyl-CoA through the involvement of 4-hydroxybutyryl-CoA dehydratase (Pearson et al. 2019).

Comparing the 3HP-4HB cycle with the Calvin-Benson-Bassham (CBB), there are important differences. The 3HP-4HB cycle involves the formation of one molecule of glyceraldehyde 3-phosphate from three molecules of CO_2, resulting in comparable energetic terms. However, the oxygenase activity of ribulose-1,5-bisphosphate carboxylase-oxygenase (RuBisCO) renders the CBB cycle energetically inferior (Loder et al. 2016). The 3HP-4HB cycle's significance goes beyond carbon dioxide fixation, as it holds potential applications in biotechnology and bioengineering. Researchers are exploring its use in developing methods to produce biofuels and other valuable bioproducts from carbon dioxide and waste streams. Moreover, scientists have engineered bacteria to more efficiently convert carbon dioxide into organic molecules for various industrial purposes.

1.4.3 Reductive Acetyl-CoA Pathway/Wood-Ljungdahl Pathway

The Wood-Ljungdahl pathway produces energy by fixing CO_2 into organic compounds in anaerobic environments. Found in various phylogenetic classes including Archaea, the Wood-Ljungdahl pathway can operate in both oxidative and reductive directions. Acetogens are used for energy conservation and autotrophic carbon assimilation. Methanogens employ it for CO_2 fixation while converting $H_2 + CO_2$ to methane to conserve energy. Acetoclastic methanogens link it to methanogenesis to generate metabolic energy. Additionally, sulfate-reducing bacteria uses it in reverse to generate metabolic energy by coupling acetate oxidation to sulfate reduction. In this pathway, CO_2 is converted into formate (HCOO-) and then into carbon monoxide (CO) by formate dehydrogenase. CO is then combined with coenzyme A (CoA) by CO dehydrogenase/acetyl-CoA synthase to form acetyl-CoA. In the second stage, acetyl-CoA is further reduced to produce organic compounds like acetate, ethanol, or butyrate, depending on the organism and its conditions. This reduction reaction is catalyzed by a series of enzymes, including acetyl-CoA decarbonylase/synthase, which removes the carbonyl group from acetyl-CoA, and methyltetrahydrofolate synthase, which provides the reducing equivalents needed for the reduction reaction (Ragsdale 2008). The pathway's ability to synthesize organic molecules holds promise for biotechnology, facilitating the production of biofuels and valuable organic compounds.

1.4.4 C_3—C_4—CAM Pathways

These are the primary carbon fixation pathway in photosynthetic organisms, such as plants and algae. The C_3 Calvin cycle uses the energy from sunlight to convert carbon dioxide into glucose and other sugars. C_4 pathway is found in certain types of plants, such as corn and sugarcane. It allows these plants to efficiently fix carbon dioxide in hot and dry conditions. CAM pathway is found in succulents that live in arid environments. It allows these plants to fix CO_2 at night and conserve water during the day. Among these, the Calvin cycle or C_3 pathway which is prominent in photosynthetic microbes is discussed in detail in Chap. 3.

These carbon fixation pathways have evolved to allow organisms to adapt to different environments and ecological niches. Understanding these pathways is important for understanding the processes of photosynthesis, primary production in ecosystems, and global carbon cycling.

1.5 Photosynthetic Pigments—Biogenesis

Photosynthetic pigments are responsible for harvesting light energy and transducing it to the photosystem complexes to generate reducing equivalents. Photosynthetic microorganisms possess different pigments which make them unique in absorbing specific light intensities in the absorption spectra (Fig. 1.4). These pigments determine the photosynthetic efficiency and the color of the microorganisms.

1.5.1 Chlorophylls

Chlorophylls are green pigments that are found in cyanobacteria, algae, and plants and constitute the major components of the photosynthetic apparatus. They have optimal light absorption at wavelengths of 450–500 and 650–700 nm and are responsible for

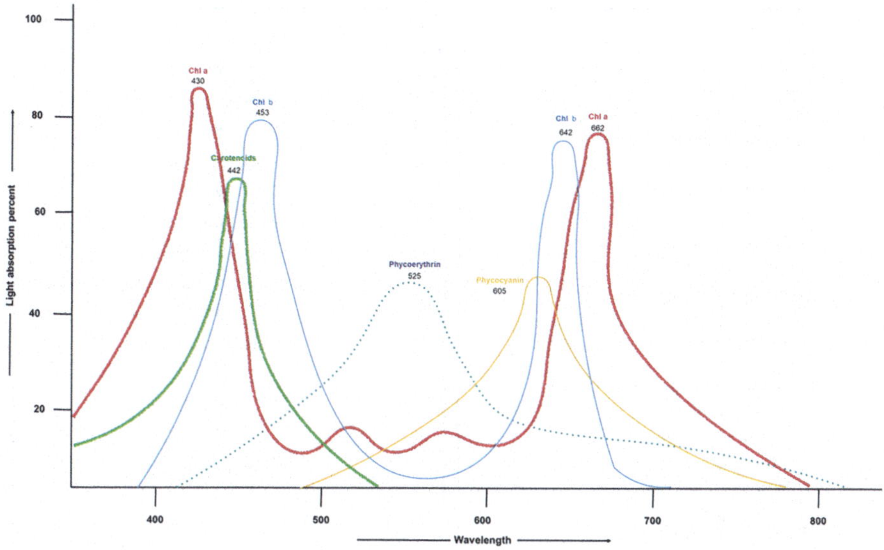

Fig. 1.4 Absorption spectra of different pigments and their wavelength maxima

the green color of the photosynthetic organisms. Chlorophyll a has a higher absorption rate than chlorophyll b wherein the former is most efficient in absorbing red and blue light, while chlorophyll b is most efficient in absorbing blue and green light. The gene sequence of Chl a monooxygenase enzyme that converts Chl a to Chl b has played a crucial role in tracing back the evolution of photosynthetic apparatus initially to green-sulfur bacteria and then branched by purple bacteria.

Chlorophyll biosynthesis is a complex process that involves multiple enzymatic steps and regulatory mechanisms (Fig. 1.5). In photosynthetic organisms, tetrapyrroles which are chelated with $Mg_2þ$ in closed rings form the backbone of chlorophyll molecules. $Fe_2þ/Fe_3þ$ chelated tetrapyrroles and open tetrapyrroles are other classes that are part of the heme family phytochromobilin, respectively. Majority of the tetrapyrroles are synthesized within plastids, with some subsequently transferred to other cell compartments. $GLU\text{-}tRNA_{Glu}$, which is also utilized for protein synthesis, is the first substrate for synthesis of tetrapyrrole within plastids. The Glu moiety gets reduced by glutamyl-tRNA reductase (Glu-R) to yield glutamic acid 1-semialdehyde (GSA), which undergoes rearrangement to yield D-aminolevulinic acid (ALA). Two molecules of ALA are condensed to porphobilinogen (PBG) which subsequently condensed to produce a linear tetrapyrrole, hydroxymethylbilane, which is subsequently cyclized into uroporphyrinogen (Uro) III. Coproporphyrinogen (Copro) III is produced when the acetic acid side chains of Uro-III are reduced to methyl groups. Protoporphyrinogen (Protogen) IX is created as a result of the reduction of the propionic acid side chains to vinyl ones, which is followed by their oxidation into protoporphyrin (Proto) IX. Mg-Proto-IX is methylated to produce Mg-Proto-IX monomethyl ester (Mg-Proto-IX-MMe) once a Mg^{2+}

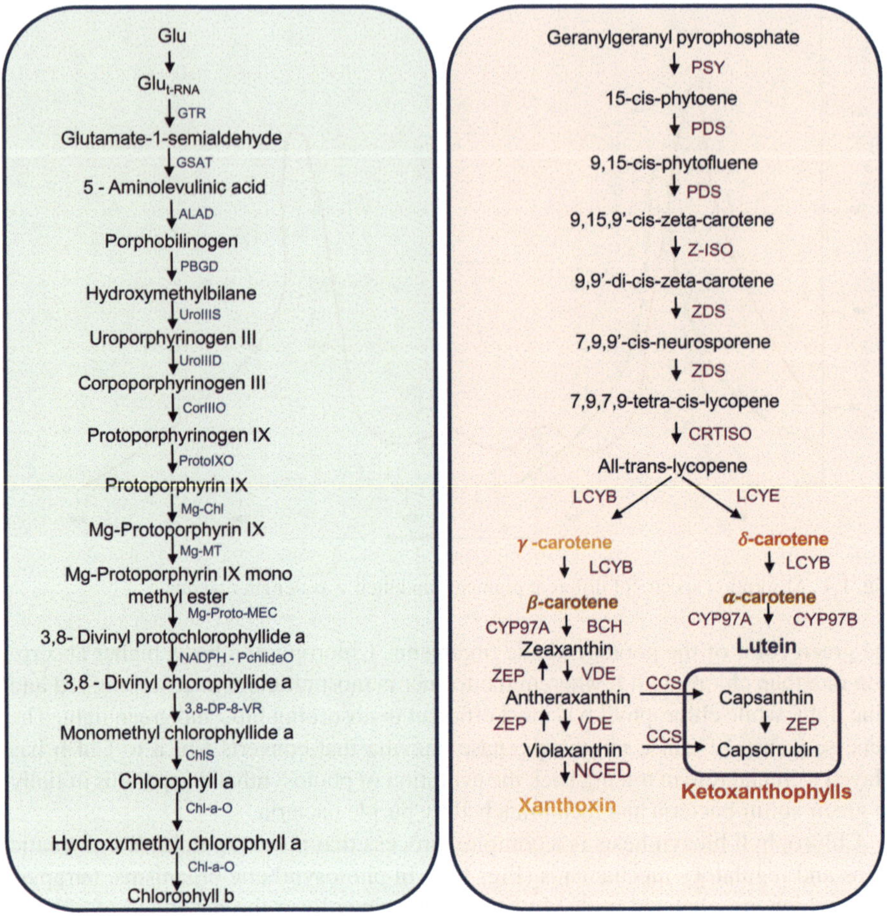

Fig. 1.5 Biosynthetic pathway of chlorophylls and carotenoids in microalgae and cyanobacteria

ion is inserted into the center of Proto-IX. The isocyclic ring is created subsequently, leading to the production of protochlorophyllide (Pchlide). Pchlide is converted to chlorophyllide (Chlide), which is then either esterified to produce chlorophyll a (Chl a) or oxidized to produce Chlide b which subsequently esterified to chlorophyll b (Chl b) (da Silva and Lombardi 2020).

1.5.2 Carotenoids

These pigments are found in plants, algae, and some bacteria majorly involved in protecting the cells from damage caused by excess light. They have optimal light absorption at wavelengths of 400–550 nm. Diatoms and brown algae possess fucoxanthin, which is a prominent carotenoid along with Chl a and c in their outer light-harvesting complexes. Fucoxanthin is a brown pigment that helps in absorbing light energy in the blue-green part of the spectrum. Carotenoids usually absorb blue and green light efficiently and appear in orange and yellow colors. Structurally, carotenoids are part of the isoprenoid family which also constitute sterols, sesquiterpenes, and triterpenoids. They are derived from isopentenyl diphosphate (IPP) and dimethylallyl pyrophosphate (DMAPP) in two independent pathways—the mevalonate (MVA) pathway and methyl erythritol 4—phosphate (MEP) pathway (Fig. 1.5). The former pathway is active in the cytosol of non-photosynthetic organisms wherein acetyl-CoA acts as a precursor for mevalonic acid through kinase reactions and decarboxylation. The key enzyme involved in regulating the generation of IPP through mevalonic acid is HMG-CoA reductase. In subsequent steps, several IPP molecules are added in a head-to-tail arrangement by the enzyme prenyltransferase to produce the terpenes and carotenoids' direct precursors, geranyl diphosphate (C_{10}), farnesyl diphosphate (C_{15}), or geranylgeranyl diphosphate (C_{20}). In photosynthetic microbes, the biosynthesis of isoprenoids follows the MEP pathway which starts from pyruvate and glyceraldehyde—3-phosphate with the involvement of ATP and NADPH. In chlorophytae algal groups, the synthesis of carotenoids, phytol, and isoprenes is achieved by DOXP/MEP pathway. The geranylgeranyl pyrophosphate is further condensed to create "phytoene" by the enzyme phytoene synthase and serves as a rate-determining step during carotenoid synthesis. Later, phytoene is converted to ζ—carotene and then to lycopene using carotenoid isomerase. The lycopene functions as a branch point where it is first transformed into β-carotene through cyclization by the action of lycopene β-cyclase and then undergoes hydroxylation reaction through carotene—β-hydroxylase to produce zeaxanthin. It is then transformed to violaxanthin by the assistance of epoxidation, which depends on oxygen and various cofactors such as ferredoxin, NADH, and FAD (Sun et al. 2016). The process of de-epoxidation, facilitated by the enzyme violaxanthin de-epoxidase, leads to the conversion of violaxanthin to zeaxanthin under conditions of high light intensity. In an alternative reaction pathway, the enzymes β-cyclase and ε-cyclases work together in a synchronized manner to produce α-carotene. This α-carotene is next subjected to hydroxylation by carotene β-hydroxylase, resulting in the formation of lutein (Sun et al. 2016).

1.5.3 Phycobilins and Fucoxanthin

Phycobiliproteins are found in cyanobacteria and red algae. They are water-soluble pigments that absorb light in the blue and green regions of the electromagnetic spectrum and subsequently transfer the energy to chlorophylls for photosynthesis. Phycobilisomes (PBS) which are protein–pigment complexes enhance light-capturing capacity of these microbes with the help of pigments such as phycocyanin and phycoerythrin. Though the presence of phycobiliproteins is common among cyanobacterial species, Prochlorophyceae is characterized by the presence of Chlb pigment but the absence of phycobiliproteins, distinguishing it from other cyanobacterial groups. Prochlorophytes that are devoid of phycobilins include *Prochlorococcus*, *Prochlorothrix*, and *Prochloron*. Red algae also exhibit transmembrane antennas known as LHCR proteins, which are connected with Photosystem I (PSI) (Bhattacharya et al. 2013; Neilson and Durnford 2010). These multimeric particles are located on the protoplasmic surface of the thylakoid membrane. Phycobilisomes are constructed in proximity to allophycocyanobilin-containing cores, which are interconnected with the cores of PS II. Phycocyanobilin can be found in the phycobiliprotein discs that surround the center of the molecule. Depending on the species, the more distal discs are made of phycoerythrobilin or phycourobilin.

1.6 Conclusion

It is important to understand how the evolutionary history of photosynthesis did, from the emergence of anoxygenic photosynthesis to the development of oxygenic photosynthesis, shape the diversity of photosynthetic organisms, their metabolic pathways, and their significant impact on Earth's atmosphere and ecosystems. These aspects could be studied in certain model species of unicellular photosynthetic microbes of *Cyanobacteria* and microalgae that perform oxygenic photosynthesis. Heliobacteria, acidobacteria, and green and purple sulfur bacteria do anoxygenic photosynthesis by using bacteriochlorophylls and carotenoids especially in low-light aquatic habitats. *Cyanobacteria* are essential to the carbon cycle and oxygen synthesis. The reverse Krebs cycle, Wood-Ljungdahl pathway, 3-hydroxypropionate-4-hydroxybutyrate cycle, 3HP-4HB pathway, and reductive acetyl-CoA pathway are crucial for carbon fixation and biosynthesis. Current understanding of these carbon fixation pathways is key to develop sustainable carbon sinks to tackle climate change. The pigment biosynthetic pathways that are conserved among the photosynthetic microbes are crucial in improving the efficiency of these microbes for industrial applications through advanced biotechnological interventions.

References

Behera B et al (2014) Sulphur oxidising bacteria in mangrove ecosystem: a review. Afr J Biotechnol 13:2897–2907

Bhattacharya D et al (2013) Genome of the red alga Porphyridium purpureum. Nat Commun 4(1):1–10

Buchanan BB, Arnon DI (1990) A reverse KREBS cycle in photosynthesis: consensus at last. Photosynth Res 24:47–53

Chakdar H, Pabbi S (2016) Cyanobacterial phycobilins: production, purification, and regulation. In: Frontier discoveries and innovations in interdisciplinary microbiology. Springer, New Delhi, pp 45–69
Cohen Y, Gurevitz M (2006) The cyanobacteria—ecology, physiology and molecular genetics. Prokaryotes 4:1074–1098
da Silva JC, Lombardi AT (2020) Chlorophylls in microalgae: occurrence, distribution, and biosynthesis. In: Pigments from microalgae handbook. Springer, Cham, pp 1–18
Dedysh SN, Damsté JSS (2018) Acidobacteria. eLS 4:1–10
Durai P et al (2015) Structure and effects of cyanobacterial lipopolysaccharides. Mar Drugs 13:4217–4230
Frain KM et al (2016) Protein translocation and thylakoid biogenesis in cyanobacteria. Biochim Biophys Acta Bioenerg 1857:266–273
Heimann K, Cirés S (2015) N_2-fixing cyanobacteria: ecology and biotechnological applications. In: Handbook of marine microalgae. Elsevier, London, pp 501–515
Heinnickel M, Golbeck JH (2007) Heliobacterial photosynthesis. Photosynth Res 92:35–53
Hoiczyk E (2000) Gliding motility in cyanobacteria: observations and possible explanations. Arch Microbiol 174:11–17
Koller M (2015) Cyanobacterial polyhydroxyalkanoate production: status quo and quo vadis? Curr Biotechnol 4:464–480
Komárek J (2016) A polyphasic approach for the taxonomy of cyanobacteria: principles and applications. Eur J Phycol 51:346–353
Krzmarzick MJ et al (2012) Natural niche for organohalide-respiring chloroflexi. Appl Environ Microbiol 78:393–401
Kumar S et al (2017) Molecular insights into photosynthesis and carbohydrate metabolism in Jatropha curcas grown under elevated CO_2 using transcriptome sequencing and assembly. Sci Rep 7(1):11066
Loder AJ et al (2016) Reaction kinetic analysis of the 3-hydroxypropionate/4-hydroxybutyrate CO_2 fixation cycle in extremely thermoacidophilic archaea. Metab Eng 38:446–463
Madhana Sekhar K et al (2014) Persistent stimulation of photosynthesis in short rotation coppice mulberry under elevated CO_2 atmosphere. J Photochem Photobiol B 137:21–30
Muchowska KB et al (2017) Metals promote sequences of the reverse Krebs cycle. Nat Ecol Evolut 1:1716–1721
Neilson JA, Durnford DG (2010) Structural and functional diversification of the light-harvesting complexes in photosynthetic eukaryotes. Photosynth Res 106:57–71
Nicolaisen K et al (2009) The outer membrane of a heterocyst-forming cyanobacterium is a permeability barrier for uptake of metabolites that are exchanged between cells. Mol Microbiol 74:58–70
Oren A (2014) Cyanobacteria: biology, ecology and evolution. In: Cyanobacteria: an economic perspective. Wiley, Hoboken, pp 1–20
Pearson A et al (2019) CO_2-dependent carbon isotope fractionation in Archaea, part I: modeling the 3HP/4HB pathway. Geochim Cosmochim Acta 261:368–382
Pfennig N, Widdel F (1982) The bacteria of the sulphur cycle. Philos Trans R Soc Lond B Biol Sci 298:433–441
Pinevich A et al (2012) Cyanobacteria of the genus prochlorothrix. Front Microbiol 3:173
Ragsdale SW (2008) Enzymology of the Wood–Ljungdahl pathway of acetogenesis. Ann N Y Acad Sci 1125:129–136
Ranadheer P et al (2019) Non-lethal nitrate supplementation enhances photosystem II efficiency in mixotrophic microalgae towards the synthesis of proteins and lipids. Bioresour Technol 283:373–377
Sciuto K, Moro I (2015) Cyanobacteria: the bright and dark sides of a charming group. Biodivers Conserv 24:711–738
Shevela D et al (2013) Oxygenic photosynthesis in cyanobacteria. CRC Press, Boca Raton
Singh RK et al (2011) Cyanobacteria: an emerging source for drug discovery. J Antibiot 64:401–412

Sirevåg R, Ormerod J (1970) Carbon dioxide fixation in green sulphur bacteria. Biochem J 120:399–408

Speirs LB et al (2019) The phylogeny, biodiversity, and ecology of the Chloroflexi in activated sludge. Front Microbiol 10:2015

Sreeharsha RV, Venkata Mohan S (2021) Symbiotic integration of bioprocesses to design a self-sustainable life supporting ecosystem in a circular economy framework. Bioresour Technol 326:124712

Sreeharsha RV et al (2019) Mitigation of drought-induced oxidative damage by enhanced carbon assimilation and an efficient antioxidative metabolism under high CO_2 environment in pigeonpea (*Cajanus cajan* L.). Photosynth Res 139(1):425–439

Sreeharsha RV et al (2023) Orienting biodiesel production towards sustainability and circularity by tailoring the feedstocks and processes. J Clean Prod:137526

Sun Z et al (2016) Microalgae as a source of lutein: chemistry, biosynthesis, and carotenogenesis. Adv Biochem Eng Biotechnol 153:37–58

Ting CS (2014) The architecture of cyanobacteria, archetypes of microbial innovation. In: The structural basis of biological energy generation. Springer, Dordrecht, pp 249–275

Trampe E, Kühl M (2016) Chlorophyll f distribution and dynamics in cyanobacterial beachrock biofilms. J Phycol 52:990–996

Waterbury JB (2006) The cyanobacteria—isolation, purification and identification. Prokaryotes 4:1053–1073

Yang SH et al (2016) Prevalence of potential nitrogen-fixing, green sulfur bacteria in the skeleton of reef-building coral Isopora palifera. Limnol Oceanogr 61(3):1078–1086

Chapter 2
Structural and Functional Dynamics of Microbial Photosystem Complexes

Abstract Photosynthesis involves two stages, the light reactions and the dark reactions that occur in thylakoid membranes and stroma, respectively. The primary function of the light reactions is to generate biochemical reductant ($NADPH_2$) and chemical energy (ATP) which are essential for the assimilation of inorganic carbon. Phototrophy is powered by Photosystem I and Photosystem II reaction centers which are classified by their absorption spectra, the order of electron transfer reactions, and terminal electron acceptors. Photosystem II plays a key role in the oxidation of water to molecular oxygen and the generation of ATP through photophosphorylation. PSII is a multimeric complex located in the thylakoid membrane, with three main components: the reaction center, the oxygen-evolving complex, and the inner light-harvesting antennae. The PSII reaction center contains D1 and D2 proteins and is responsible for the charge separation and stabilization. The PSI consists of P700, A0, A1, and iron-sulfur clusters F_X, F_A, and F_B, which ultimately transfer electrons to ferredoxin (Fd) and ferredoxin-NADP+ reductase (FNR). The extrinsic and low molecular weight proteins in photosystem complexes work together to optimize the transfer of energy and electrons, stabilize the complex, and protect it from damage. Additionally, mutations or alterations in these proteins can lead to reduced photosynthetic efficiency and negatively impact growth and survival. Understanding the function and interactions of these proteins is important for improving yields and developing more efficient photosynthetic cell factories for various industrial applications.

Keywords Photosystem I · Photosystem II · Oxygen evolving complex · Electron transport chain · Ferredoxin · Cytochromes

2.1 Introduction

Microbial photosynthesis contributes majorly to Earth's carbon cycle and oxygenation of the atmosphere. The phenomenon of capturing and transforming light energy into chemical energy has been fascinating the scientists for several decades

R. V. Sreeharsha, S. Venkata Mohan, *Microbial Photosynthesis*,
https://doi.org/10.1007/978-981-97-1253-3_2

and significant contributions have been made in understanding the molecular mechanisms underlying the obscure photochemical reactions. Since the Devonian era, the thylakoid membranes and light-harvesting accessory pigments have been evolving continuously nevertheless, the fundamental mechanisms of photochemistry involving electron transfer between photosystems, synthesis of energy equivalents, and carbon assimilation are conserved across phyla of photosynthetic organisms. Of late, the exploration of mechanistic aspects of photosynthetic reactions paved the way to design microbial photosynthetic cell factories for environmental engineering and renewable energy production with commercial applications (Sreeharsha and Venkata Mohan 2021; Sreeharsha et al. 2023).

In all photosynthetic microbes, the fundamental aspect of oxygenic photosynthesis is to convert carbon dioxide and water into oxygen and carbohydrates. To understand this fundamental concept of photolysis, it is crucial to comprehend the structural organization of the chloroplasts—the most conserved green cell organelle of photosynthetic organisms. Over the years, the chloroplasts have evolved into specialized cell organelles that compartmentalize different steps of light and dark reactions. The light reactions that take place in the presence of light energy occur in the thylakoid membranes of the chloroplast and dark reactions that do not require light energy occur in the stroma of chloroplast. The two cellular events are interconnected to each other through the exchange of redox energy and feedback inhibition. Electron microscopy of chloroplast revealed the structural features of thylakoid membranes and stroma which helps in understanding the functional aspects of the photosystems in detail. In *Chlamydomonas reinhardtii*, chloroplast occupies approximately 40% of the total cell volume. The chloroplast genome is highly conserved with a majority of genes related to light-dependent activities of photosynthesis or involved in transcription and translation. Primarily, thylakoids consist of intra-thylakoid space, lumen, and its surrounding membrane with glycolipid bilayers of galactosylglycerol moieties. Numerous complexes involving protein–protein or pigment–protein interactions are distributed across the lipid bilayer membrane of the thylakoid. In some instances, proteins may extend partly into the bilayer, with certain functional groups oriented towards the lumen or stroma. The adjacent thylakoids which are organized in pairs or stacks of three are interconnected with each other through a single-layered photosynthetic lamella. The lamellae seen in cyanobacteria and algae are comprised of hydrophilic phycobilisomes, which function as external complexes for the absorption of light. In contrast, the chloroplasts seen in higher plants have distinct zones of densely packed thylakoids known as grana, which are interconnected by stromal lamellae.

There are structural differences in the arrangement of thylakoids and photosystem complexes in cyanobacteria and microalgae. Photosystems of photosynthetic bacteria are structurally and functionally similar to other photosynthetic organisms; however, they are localized to the plasma membrane. Cyanobacteria have larger antenna complexes compared to microalgae, which is believed to result in a higher rate of light absorption and more efficient energy transfer. The exact mechanisms of energy conversion and water splitting may differ between microalgae and cyanobacteria due to differences in the composition and arrangement of the pigments and electron transfer pathways. Microalgae have developed mechanisms to cope with fluctuations in light intensity and temperature, while cyanobacteria have evolved to live in low-light environments.

The functional dynamics of thylakoid membranes are determined by five major protein complexes: light-harvesting antennae, Photosystem II (PSII), Photosystem I (PSI), cytochrome b6/f, and ATP synthase, which maintain photosynthetic electron transport and photophosphorylation (Fig. 2.1a). The two reaction centers PSI and PSII are differentiated based on their absorption spectra, order of electron transfer reactions, and terminal electron acceptors. Though it is common to have two photosystems for electron transport in all photosynthetic microbes, anoxygenic phototrophs use one reaction center either PSI or PSII. The operation of photosystems occurs in a sequential manner, with each photosystem being linked together by a

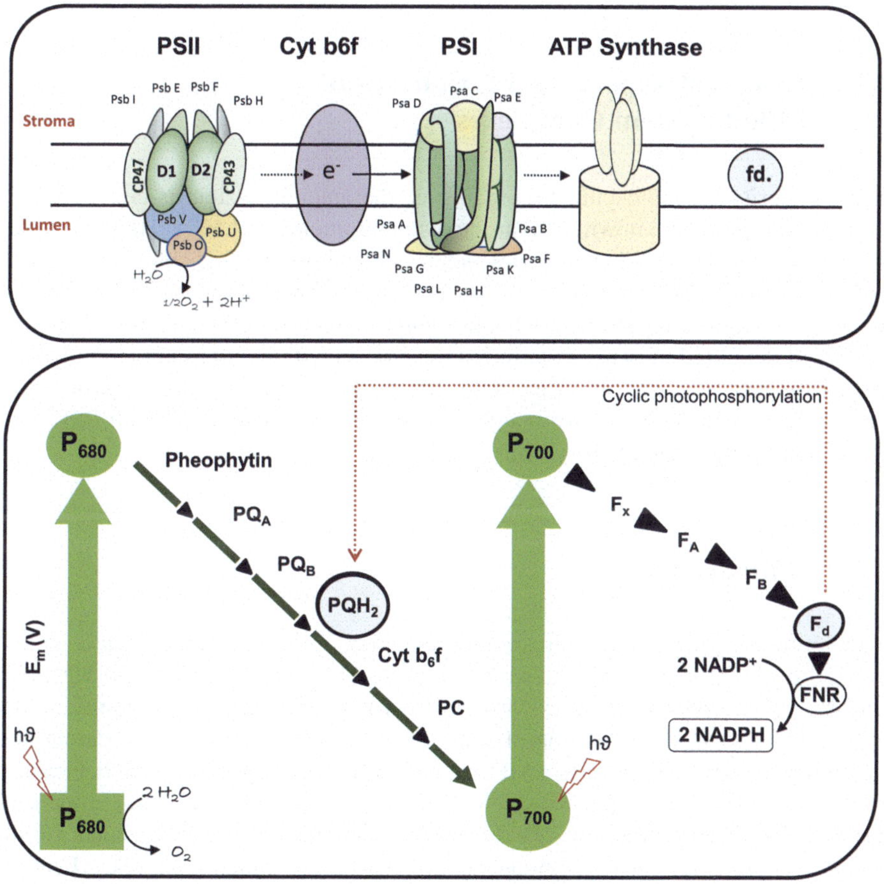

Fig. 2.1 Diagrammatic representation of photosystem complexes and electron flow in photosystems in microalgae and cyanobacteria. Represented the intrinsic and extrinsic proteins of PSII and PSI along with the Z-scheme of the electron transport in photosystems

chain of electron carriers. This process is elucidated most effectively through the use of the "Z" scheme, which serves as a depiction of the transit of electrons in photosynthesis (Hill and Bendall 1960) (Fig. 2.1b). According to this, the redox components are characterized by their equilibrium midpoint potentials and the electron transport reactions proceed energetically downhill, from a lower (more negative) to a higher (more positive) redox potential. Theoretically, the energy of at least eight photons is required to carry out this process (Blankenship 2021).

This chapter gives a comprehensive understanding of the components of the photosystem complexes and their mechanistic aspects. The protein—pigment complexes of light-harvesting complexes and the state transitions of photosystem complexes are discussed in detail. The effect of various environmental conditions on the functioning of photosystems and the strategies to cope with stress conditions are given in detail.

2.2 Structural Aspects and Localization of Photosystem Complexes

In the majority of photosynthetic organisms, the antenna system is comprised of hydrophobic proteins known as light-harvesting complexes (LHC). These proteins are found in the thylakoid membranes and have the ability to bind chlorophyll (Chl) and carotenoids (Car). Photosystem I (PSI) is situated within the stacked thylakoid membranes known as stroma, whereas Photosystem II (PSII) is found in the unstacked thylakoid membranes referred to as grana within chloroplasts (Anderson and Chow 2002; Apostolova et al. 2006; Mustardy et al. 2008). Along with light-harvesting complexes there are many extrinsic proteins associated with PSII and PSI which stabilize the photosystems (Table 2.1).

2.2.1 Photosystem II

Photosystem II (P680) is a multi-subunit membrane protein complex that contains over 20 different subunits, with a relative molecular mass of about 350 kDa and absorbs light at 680 nm. Along with the light-harvesting antenna complex, PSII plays a key role in the oxidation of water to molecular oxygen thus initiating the electron transport chain. The PSII isolated from thermophilic cyanobacterium *Thermosynechococcus vulcanus* was studied extensively for its structural properties through X-ray crystallography. This provided a model for the arrangement of all energy and electron transfer components in other photosynthetic microbes (Umena et al. 2011). The PSII complex is a culmination of different protein and pigment components including chlorophyll a, chlorophyll b, carotenoids, and reaction center—P680, oxygen-evolving complex, inner light-harvesting antennae, electron

Table 2.1 Extrinsic proteins of PSI and PSII light-harvesting complexes

Extrinsic protein	Corresponding gene	Mol weight (KDa)	Function
Photosystem II			
PsbA	psbA	24	Protecting PSII from oxidative damage in higher plants
PsbB	psbB	20.20	Chlorophyll binding; mRNA binding; binds at Mn center
PsbC	psbC	43	It binds chlorophyll and helps catalyze the primary light-induced photochemical processes of PSII
PsbD	psbD	39.50	Provides stability in the membrane of the complex and regulates the expression of the D1 protein of the complex
PsbE	psbE	9.30	Lyase activity
PsbF	psbF	4.40	Helps to stabilize and facilitate the function of OEC in PSII
PsbH	psbH	7.70	PsbH is important for prompt incorporation of the newly synthesized D1 protein into PSII complexes and for the past D1 maturation
PsbI	PsbI	4.10	It is early assembly partner of the D1 protein, and it plays a functional role in stabilizing the binding of CP43 in the PSII holoenzyme
PsbO	PsbO	33	Stabilizes the Mn_4CaO_5 cluster to the suboptimal Cl concentration and protects the cluster from exogenous reductants
PsbP	psabP	23	This protein regulates the binding properties of Ca^{+2} and Cl^-; stabilizes the Mn cluster of PSII
PsbQ	psbQ	17	Participates in normal function of photosynthetic water oxidation; it acts as an accessory subunit to support PsbP in higher plants; PsbQ is shown to be essential for photoautotrophic growth under conditions of limited Ca^{+2}, Cl^-, and Fe
Photosystem I			
PsaA	psaA	81.70	Part of an ABC transporter complex involved in Mn import; related large transmembrane protein involved in binding of P700, A0, A1, Fx part of the photosynthetic reaction center protein family
PsaB	psaB	81.40	Same as PsaA function
PsaC	psaC	10	Provides an essential lysine residue for fast electron transfer to ferredoxin
PsaD	psaD	15.50	Provides docking site for ferredoxin and interact with the PsaB, PsaC, and PsaL subunit of PSI
PsaE	psaE	8.40	Involved in docking of ferredoxin/flavodoxin to the PSI complex and also participates in the cyclic electron transfer around PSI
PsaF	psaF		It functions as a pH sensor and indicates that the presence of PsaF is important for PsaE stability
PsaG	psaG		Binding of Lhc a1/4; regulation of PS1 (6.32)

(continued)

Table 2.1 (continued)

Extrinsic protein	Corresponding gene	Mol weight (KDa)	Function
PsaH	psaH		Binding of LHC II (state transition); stabilization of PsaD
PsaI	psaI	4	Stabilizes the light-harvesting complex to docking site of the PSI
PsaK	psaK	3	In stabilization of the protein—pigment complex
PsaL	psaL		Assembly and stabilization of PSI; it involved in facilitating the transfer of electron within the PSI complex; protection from oxidative damage
PsaN	psaN		Involved in interaction of plastocyanin and PSI complex; involved in the docking of plastocyanin

donor—pheophytin and primary and secondary acceptors—QuinoneA (Q_A), QuinoneB (Q_B) as well as various other proteins that play a crucial role in stabilization and transport. On the stromal side of the thylakoid membrane, the PSII components interact with each other to initiate and accomplish the electron transport chain (Fig. 2.1).

The biogenesis of PSII is a highly regulated and complex process that is essential for the efficient capture of light energy and the production of oxygen during photosynthesis. It involves the synthesis and assembly of PSII subunits, insertion of cofactors, and final membrane integration. The genes encoding PSII subunits are expressed into precursor proteins and directed through N-terminal targeted proteins into chloroplasts where they are assembled as pre-PSII complexes in the thylakoid membranes (Anderson et al. 2002). These pre-PSII complexes are composed of PSII subunits, including the D1 and D2 proteins, which are the core components of the PSII reaction center (van Amerongen and Croce 2013). The pre-PSII complexes undergo maturation by incorporating various cofactors, including chlorophylls and carotenoids, pheophytins, metal ions, and additional subunits to form functional PSII. These cofactors are essential for capturing and transferring light energy within the PSII complex. The fully assembled PSII complex including D1 and D2 proteins, α and β subunits of cyt b559 complex, and chlorophylls and pheophytins integrate into the thylakoid membrane, where it is embedded in a matrix of lipids and other membrane proteins. Among other proteins of the PSII complex, D1 and D2 proteins form the central core and are equipped with all essential prosthetic groups necessary for charge separation and stabilization. The intrinsic Chl *a* proteins CP47 and CP43 are flanked on either side of the D1–D2 heterodimer and contribute majorly to the transfer of excitation energy to the reaction center. They also stabilize other extrinsic proteins on the luminal side wherein CP43 specifically contributes an amino acid ligand to the oxygen-evolving complex (OEC). The OEC cluster with four manganese ions and a calcium ion (Mn_4CaO_5) is involved in splitting water molecules into oxygen, protons, and electrons thus releasing oxygen into the atmosphere as a by-product (Nickelsen and Rengstl 2013) (Refer Sect. 2.3.1).

The maturation process is tightly regulated to ensure the correct assembly of PSII and to prevent the accumulation of non-functional or misfolded complexes. Once PSII is fully assembled and functional, it is maintained through a variety of mechanisms, including protein turnover, repair of damaged components, and regulation of PSII activity in response to changing environmental conditions. Any disruption in this process can result in defects in PSII function and photosynthesis, leading to decreased growth and productivity. PSII acts as a marker for assessing growth, and photosynthetic physiology as it is very sensitive to changes in environmental and growth variables including pH, temperature, and nutrient status. The assessment of PSII kinetics in microalgae and cyanobacteria will provide their level of tolerance to various abiotic stress conditions. Hence, PSII is an important cue for screening efficient microbial strains for higher growth and biomass productivity (Faseela et al. 2020; Reddy et al. 2019; Gokhale and Sayre 2009; Madhana Sekhar et al. 2014; Ranadheer et al. 2019; Sekhar et al. 2015; Sharma et al. 2020).

2.2.2 *Photosystem I*

Photosystem I (P700) absorb light with longer wavelengths than PSII usually at 700 nm and transfers its high-energy electrons to a series of electron carriers. Photosystem I (PSI) is a multi-subunit, thylakoid membrane protein consisting of more than 110 subunits, with a molecular weight of about 1 megadalton. It is composed of a core complex containing reaction center, chlorophyll a, chlorophyll b, and carotenoids 12 transmembrane helices, and several peripheral subunits (Fig. 2.1). The peripheral subunits are involved in electron transfer and stabilization of the core complex whereas, the core complex is responsible for capturing light energy. *PsaA* and *PsaB* are the other two major proteins that embed into the membrane aligning their prosthetic groups to stroma (Table 2.1). Furthermore, the membrane also contains embedded components such as Chl dimer P700, which is responsible for initiating primary charge separation, as well as electron carriers A_0 (Chla), A1 (phylloquinone), and F_X (4Fe–4S). Generated electrons are further transported to the 4Fe–4S electron acceptors F_A and F_B of the *PsaC* subunit and to the terminal mobile electron acceptor, which is ferredoxin. The high-resolution structure of PSI is known for both cyanobacteria and microalgae (Amunts et al. 2010; Jordan et al. 2001). PSI also involves cyclic electron flow, a process in which electrons cycle between PSI and the electron transport chain, producing additional ATP molecules without the production of NADPH.

2.2.3 Cytochrome Complexes and Its Component

Cytochrome complexes are a family of proteins that are involved in electron transport in cells. In general, they are located in the inner mitochondrial membrane as well as inner thylakoid membrane of chloroplast and participate in the electron transport chain, which generates ATP, the energy currency of cells. The cytochrome complexes are composed of several components, including heme groups, iron-sulfur clusters, and various protein subunits. The heme groups are responsible for accepting and donating electrons, while the iron-sulfur clusters serve as electron carriers. The protein subunits provide structural support and help to regulate the electron transport process. Concerning chloroplasts, cytochrome complex is an essential component of the light reactions of photosynthesis. They are embedded within the thylakoid membrane and play a critical role in the transfer of electrons from PSII to PSI which is essential for the generation of ATP and NADPH. Cytochrome b6f protein complex which is embedded in the thylakoid membrane accepts electrons from the water-splitting complex in PSII (Fig. 2.1). These electrons are then transferred through a series of electron carriers including cytochrome *f*. As electrons move through the complex, they release energy, which is used to pump protons (H^+ ions) from the stroma into the thylakoid space, creating a proton gradient. The proton gradient is crucial for the generation of ATP. Protons flow back into the stroma through ATP synthase, driving the synthesis of ATP from ADP and inorganic phosphate (Pi) in a process called chemiosmotic phosphorylation. Cytochrome *f* acts as a carrier for electrons between the b-type cytochromes and the soluble electron carrier plastocyanin. By transferring electrons between these carriers, cytochrome *f* helps in maintaining the flow of electrons through the complex, ultimately contributing to the generation of reducing equivalents.

2.2.4 ATP Synthase/ATPase

ATP synthase is an enzymatic complex that is embedded inside the cellular membrane. It consists of two subunits, CF_0 and CF_1. The CF_0 subcomplex has a relative molecular mass ranging from 110 to 160 kDa, while the CF_1 subcomplex has a relative molecular mass of around 400 kDa. The enzymatic complex, driven by the pH gradient, facilitates the catalytic reaction that leads to the formation of ATP from ADP and Pi. The CF_0 protein, which exhibits hydrophobic properties, is responsible for spanning the thylakoid membrane. On the other hand, the CF1 protein, which is hydrophilic in nature, is connected to CF_0 on the stromal side of the membrane. The subunit CF_0 functions as a proton channel, facilitating the movement of protons. This proton flux then energizes the subunit CF_1, which assumes a ring-like configuration and contains catalytic sites responsible for ATP production. The generation of one ATP molecule necessitates the passage of about four protons (Boyer 2002; Nelson and Junge 2015).

2.3 Functional Dynamics of Photosystem Complexes

The functional dynamics of photosystem complexes involve a series of complex processes that are crucial for the efficient and effective functioning of photosynthetic machinery. The schematic events include light absorption, electron transfer, generation of proton gradient across the thylakoid membrane, generation of ATP from ADP and inorganic phosphate by ATP synthase enzyme, and reduction of $NADP^+$ to NADPH. Light-harvesting complexes play a versatile role in the functional dynamics of photosystems by altering their structural chemistry in response to light absorption (Kumar et al. 2017). The functional dynamics of major events are discussed in the following sections.

2.3.1 Water Splitting and Oxygen Evolution—Photosystem Performance Indicators

According to estimates, photosynthesis is believed to contribute to the production of around 70% of the oxygen present in the Earth's atmosphere, rendering it a very significant biological activity of utmost importance on a global scale. Water splitting and oxygen evolution are processes that occur during photosynthesis, which is the basis for the electron transport chain and carbon fixation. The light energy absorbed by chlorophyll pigments (PSII) is used to split water molecules thereby releasing O_2 gas as a by-product. The resulting electrons from the water-splitting reaction are used to power the formation of ATP (adenosine triphosphate), which is the primary energy source of cells. This reaction is catalyzed by a complex of proteins and pigments known as the oxygen-evolving complex (OEC), which is located within the thylakoid membrane. The OEC consists of four manganese ions and a calcium ion, which work together to extract electrons from water molecules and produce oxygen gas. Calcium serves as an essential cofactor for the OEC and its depletion leads to the total impairment of the water oxidation capacity of PSII. Strontium is the only element capable of functionally substituting calcium in this context (Koua et al. 2013).

The P680+• species possesses a significantly higher redox potential, enabling it to acquire an electron from the secondary electron donor, TyrZ. This electron transfer event results in the generation of a neutral radical, TyrZ•, which subsequently facilitates the rapid oxidation of water at the oxygen-evolving center (OEC) located on the donor side, occurring within milliseconds (Lubitz et al. 2019; Pérez-Navarro et al. 2016; Yano and Yachandra 2014). During the process of water splitting and oxygen evolution, the OEC undergoes through a series of five distinct redox states, denoted as S0 to S4. These redox states are achieved by altering the oxidation valences of the four manganese ions present in the OEC. The S0 state represents the initial and most reduced state, while the S1 state is characterized as the dark stable state. The metastable S2 and S3 states undergo eventual decay to the stable S1 state, which lacks luminescence. On the other hand, the S4 state is a transitory state that liberates dioxygen and then decays to the S0 state (Krewald et al. 2015; Yano and Yachandra 2014; Dau and Haumann 2007).

The NADPH generated at the acceptor side of PSI forms the central hub for various metabolisms including the Calvin cycle, nitrate metabolism, lipids, amino acids, and pigments metabolisms. In case of higher plants 3 ATP and 2 NADPH are utilized for CO_2 assimilation and most likely the ratio of ATP/NADPH ratio associated with photosynthetic electron transport in vivo is likely below 1.5 (Allen 2002). Hence, it is postulated that photosynthetic organisms divert a portion of their electrons towards alternate acceptors apart from NADP+, in order to generate a proton gradient facilitating ATP synthesis, but without yielding any net NADPH production. While alternate sinks for electron flow generally account for a modest percentage (about 10%) of the overall linear electron flow, they play a crucial role in maximizing CO_2 absorption. Alternative electron flows in plants and microalgae encompass cyclic electron flow (CEF) around Photosystem I (PSI) and the water-to-water cycles (WWCs) (Shikanai 2007). WWCs encompass two distinct processes: the photoreduction of O_2 by Photosystem I (PSI), known as the *Mehler reaction*, and the non-photochemical oxidation of plastoquinol downstream of Photosystem II (PSII), facilitated by the plastoquinone (plastid) terminal oxidase (PTOX) reaction (Zehr and Kudela 2009). Photorespiration, which refers to the oxygenase activity of Rubisco, might be seen as a form of WWC (Ort and Baker 2002). In due course, the movement of reducing equivalents from the chloroplast to the mitochondrial oxidases, facilitated by the malate shuttle, furthermore results in the reduction of molecular oxygen by electrons produced by PSII. Consequently, this process may be seen as a form of WWC.

2.3.2 Charge Separation in Photosystem Complexes as a Function of Incident Light

When a light-harvesting complex or reaction center absorbs a photon of light or receives an exciton of energy from neighboring light-gathering antennae, it undergoes a charge separation event thus creating a proton gradient across the thylakoid membrane. This results in producing a high-energy compound ATP through photophosphorylation and a biochemical reductant NADPH which will be further utilized for the assimilation of inorganic carbon. PSI reduces ferredoxin as the terminal electron acceptor, whereas PSII reduces quinones. The transfer of electrons through the reaction center of PSII is a highly organized process, involving a series of electron transfer reactions that are catalyzed by the protein subunits. The first step involves the transfer of an electron from the excited chlorophyll molecule to a primary electron acceptor, which is a molecule of pheophytin linked to the D1 protein. This transfer results in the reduction of pheophytin, which then passes the electron on to a secondary electron acceptor, Q_A, the molecule of plastoquinone located within the PSII complex (Shevela et al. 2021). Plastoquinone is a mobile electron carrier that moves through the thylakoid membrane, transferring electrons to a series of other electron carriers, including cytochrome b6f and plastocyanin. When a reaction center with a positive charge and a Q_A with a negative charge combine, two chemical events take place. Firstly, the oxidation of water to molecular oxygen takes place after four rotations of PSII. Secondly, the reduction of plastoquinone to plastohydroquinone occurs after two turnovers of PSII (Mamedov et al. 2015). The plastoquinone molecule, after being reduced, exits PSII and becomes part of the plastoquinone pool located in the membrane. It is worth mentioning that the Cyt b6f complex has the ability to accept two electrons from the PQ pool. However, it subsequently transfers these electrons to plastocyanine (PC) in a one-electron mode. Also, these carriers are organized in a series of redox reactions, which generate the proton gradient across the thylakoid membrane. The proton gradient is then used to drive ATP synthesis through ATP synthase, which converts the energy stored in the proton gradient into the chemical energy of ATP.

Similar to Photosystem II (PSII), Photosystem I (PSI) also absorb photons, resulting in charge separation within PSI. Subsequently, the PSI reaction center becomes prepared to receive the electron from the plastocyanin (PC) in a univalent manner. The electron transfer chain of PSI comprises several components, including P700 as the principal electron donor, A0 which consists of a pair of chlorophyll molecules, A1 composed of phylloquinone molecules, and iron-sulfur clusters F_x, F_A, and F_B (Mamedov et al. 2015). The final stage involves the transmission of electrons to ferredoxin (Fd), which functions as a central intermediary for transferring electrons to either ferredoxin-NADP+ reductase (FNR) in conventional photosynthesis or hydrogenase in hydrogen generation. It is important to note that a total of 13 ferredoxin genes have been discovered in the genome of *C. reinhardtii*. However, it is worth noting that only six of these genes, namely PETF and FDX2-6, have undergone extensive characterization (Sawyer and Winkler 2017). PETF can also transfer electrons to a large number of other proteins. The Protein Electron Transport Factor (PETF) has the ability to facilitate electron transport to a diverse range of proteins. The physiologically and biotechnologically significant [FeFe]-hydrogenases HYDA1-2, which facilitate the generation of molecular hydrogen (H_2) from protons and electrons, serve as the electron acceptor in green microalgae.

2.3.3 Electron Transport Rate and Heat Dissipation—Crucial Parameters for PS Efficacy

Photosynthesis is reliant upon the fundamental processes of energy capture and electron transfer reactions. These reactions exhibit redox chemistry with high potential and can result in photo-oxidative damage to the photosynthetic apparatus. In order to mitigate the adverse effects of these deleterious side reactions, the energy transfer processes are subject to dynamic regulation. However, it should be noted that the level of control exerted on chloroplasts is not absolute since these organelles also require a comprehensive antioxidant system to effectively manage the reactive chemicals they produce (Sreeharsha et al. 2019).

The occurrence of cyclic electron flow in *Dixoniella giordanoi* is enhanced when exposed to intense light. The transfer of electrons between Photosystem II (PS II) and Photosystem I (PS I) is facilitated by the cytochrome b6/f complex and aided by two distinct types of mobile carriers. Plastoquinones, which are lipophilic benzoquinones containing an isoprenoid chain, function as two-electron carriers connecting photosystem II (PS II) with cytochrome b6/f complexes. The regulation of the quantity of active photosystem II (PSII) centers serves as a fundamental mechanism for controlling the photosynthetic electron transport chain. The photoinhibition-repair cycle also regulates the activity of active PSII centers effectively, hence preventing permanent damage to PSI. Additionally, this mechanism likely enables the oxidized PSI centers to act as quenchers of excessive excitation energy.

Photosynthetic microorganisms experience varied light intensities when grown in a natural environment. Photosynthetic machinery especially light reactions have to adapt to the changes in the environment. The adaptation can be either by modulating light-harvesting capacity or by changing the number of active reaction centers of PSII. Pigmentation is also an important factor that shows dynamic regulation in response to light. Microalgae enhances its pigmentation in response to low-light conditions and vice versa to balance photosynthesis. However, the change in the pigmentation is not as dynamic as compared to the changes in daylight fluctuations. Hence, the instant regulation can be achieved by the pH gradient which results in increased non-photochemical quenching thus reducing the amount of energy utilized in photochemistry (Falkowski and Raven 2013; Sreeharsha and Reddy 2015). The change in pH gradient results in the conversion of violaxanthin into zeaxanthin which is a reversible reaction and directly proportional to the NPQ. When compared to higher plants, zeaxanthin-dependent NPQ was minor in microalgae. Nevertheless, the formation of diatoxanthin from monoepoxide diadinoxanthin plays a similar role as of zeaxanthin which signals thermal dissipation in major microalgae phyla *Chrysophyceae* and *Phaeophyceae* (Arsalane et al. 1994; Lavaud et al. 2004). Apart from the heat dissipation, inactivation of PSII by reducing the number of reaction centers is also an emergency acclimation of the photosystem which is characterized by an exponential decline of variable fluorescence Fv. Several studies have proven that the D1 protein of PSII will act as a suicidal protein to prevent the damage of other PSII reaction centers against oxidative damage under excess light.

2.3.4 State Transitions in Photosystems

State transition is a photoprotective mechanism observed in microalgae and cyanobacteria when they are subjected to changes in light quality or quantity. The term "dynamic redistribution" pertains to the process of reallocating light-harvesting complexes (LHCs) and excitation energy between Photosystem I (PSI) and Photosystem

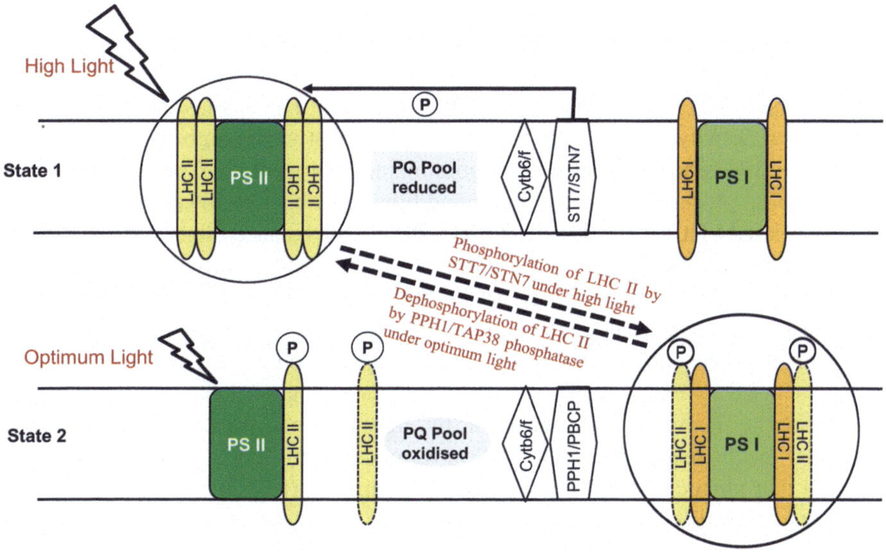

Fig. 2.2 Illustration of state transitions in photosystems in microalgae

II (PSII) in order to optimize their relative activity in response to fluctuations in light conditions (Fig. 2.2). The primary goal of state transition is to optimize light absorption and minimize photodamage under varying light intensities. State transitions are initiated by alterations in the redox state of the plastoquinone pool and manifest uniformly across the cellular environment. Fluorescence lifetime imaging microscopy (FLIM) of *Synechococcus elongatus* revealed that PSII quenching enhanced with a simultaneous decrease in PSII fluorescence in State II (Bhatti et al. 2021). In higher light conditions, the balance between light absorption by PSI and PSII becomes imbalanced, leading to an excess of absorbed energy that could potentially cause photodamage. To prevent this, the cell triggers state transition, which redistributes the LHCs from PSII to PSI or vice versa, depending on the specific conditions (Fig. 2.2). Under balanced light conditions, LHCs are evenly distributed between PSI and PSII. Both photosystems absorb light efficiently, and photosynthesis proceeds normally. When the light intensity increases, there is a higher excitation rate at PSII than at PSI, resulting in an imbalance. The excess energy absorbed by PSII can lead to the generation of reactive oxygen species (ROS) and photoinhibition, which could damage the photosystem and other cellular components. To re-establish a balance between PSI and PSII, a specific protein kinase phosphorylates the LHCs associated with PSII. The state transitions in microalgae are tightly regulated by oxidation-reduction coupled signaling which signals the redox state of plastoquinone thus changing the stoichiometry of chloroplasts (Puthiyaveetil et al. 2012). This phosphorylation event triggers state transition by causing the mobile LHCs to detach from PSII. The phosphorylated LHCs migrate from PSII to PSI which is facilitated by the association of the phosphorylated LHCs with a specific protein called the light-harvesting complex kinase (LHC II kinase STN7/STT7) protein. With the LHCs now associated with PSI, the absorption of light by PSI increases, improving its photochemical efficiency. This redistribution of LHCs helps to reduce the excess excitation energy at PSII, thus reducing the risk of photodamage and ROS production. The increased activity of PSI and reduced activity of PSII help the microalgae adapt to the high light conditions, ensuring optimal photosynthesis while minimizing photodamage. The state transition is a reversible process, and when light conditions shift back to lower intensities, the phosphorylated LHCs move back to PSII, restoring the balanced light-harvesting distribution between the two photosystems. PP2C-type phosphatase TAP38/PPH1 is responsible for dephosphorylating LHC II during State I transition. This regulatory mechanism allows microalgae to acclimate rapidly to varying light conditions, optimizing their photosynthetic efficiency, and protects the photosystems from potential damage caused by excessive light energy. Similar to higher plants and microalgae, cyanobacteria mediate state transition through phycobilisome nevertheless, the molecular mechanisms behind the PBS movement are still under characterized. It is believed that phytochromes and cyanobacteriochromes (CBCRs) play an important role in state transition in cyanobacteria in terms of two acclimation responses: far-red light photoacclimation (FaRLiP) and complementary chromatic acclimation (CCA). FaRLiP involves changes in antenna pigments to optimize light absorption under low-light conditions, while CCA allows for efficient utilization of different wavelengths of light through modification of phycobilisome composition. The changes in the fluorescence emission spectra upon the detachment of PBS from the dimeric PSII complex reveal the state transition which gets reversed by the dissociation of PBS from the monomeric PSI corresponding to ferredoxin oxidation as electron carrier and the increase of cyclic electron transport in the pigment apparatus in State I. State transitions contribute to maintaining linear electron flow during imbalanced light absorption by photosystems. State transitions can be considered as a post-translational regulation where reversible protein phosphorylation controls the distribution of excitation energy between photosystems I and II. These state transitions enable microalgae to adapt to changes in light conditions and optimize photosynthetic efficiency.

2.4 Effect of Environmental and Growth Conditions on PS Functioning

Several environmental and genetic factors influence the rate at which the photosystem electron transport takes place. Light intensity is one of the major variables that affects the reaction kinetics of the photosystem. Up to a certain point, the rate of photosynthesis rises along with light intensity and then gets saturated wherein further increase in light intensity have little impact on the reaction rate. The absorbed light at this point is known as the light saturation point. PSII and PSI have divergent coping mechanisms for photo-oxidative stress. Since the rate of damage is linearly correlated with the photon fluence rate, PSII is extremely photosensitive. However, the healing mechanisms work effectively, and the damage is quickly fixed. The inhibition of PSII activity occurs exclusively when the rate of impairment surpasses the rate of healing. The rate of repair is subject to dynamic modulation by environmental cues and the energy state of the photosynthetic apparatus. In situations of excessive light, the repair process decelerates due to feedback inhibition from dark reactions, which compromises the capacity to absorb energy from light reactions. The PSI centers exhibit a high degree of photoprotection compared to PSII. In the event of damage, the recovery of PSI is characterized by a remarkably slow process. However, strong light illumination poses a serious threat to PSI if PSII feeds the electron transfer chain (ETC) with more electrons than its electron acceptors on the reducing side can handle. When the transfer of electrons from Photosystem II (PSII) to Photosystem I (PSI) is appropriately controlled, such as in the presence of DCMU, PSI not only protects itself against photo-oxidative stress but also efficiently dissipates excess excitation energy gained by the light-harvesting system.

The reaction kinetics of the photosystem are also influenced by temperature. The ideal temperature for photosynthesis varies depending on the species, but for the majority of photosynthetic organisms, temperatures between 20 and 30 °C are thought to be ideal. The rate of reaction reduces as temperatures rise over this range as the pigments and other proteins in the reaction center get denatured. The reaction rate also declines below the ideal range when the metabolic processes necessary for photosynthesis become more sluggish. Heavy metals like lead and mercury, as well as herbicides, can attach to the reaction center and obstruct the flow of electrons thus reducing the photosynthetic efficiency and reaction rates. The presence of salt in cyanobacteria can impact the reaction kinetics, which can change the water potential of the thylakoid membranes. The reaction kinetics of PSII in cyanobacteria can also be affected by genetic variables, such as differences in the genes encoding the reaction center components. In addition to reducing the photosynthetic activity, the environmental factors and nutrients have a significant impact on photosystems by producing a variety of reactive oxygen species (ROS) (Murata et al. 2007; Szabó et al. 2005). Additionally, during redox processes, the oxygen produced from OEC also has a propensity to produce ROS which eventually results in oxidative damage (photoinhibition). High dissolved oxygen concentrations, reflection, predation, process hindrances are among the factors that affect efficiency in outdoor cultures of photosynthetic microbes (Melis 2009). Additionally, photobioreactor performance, fouling, and culture maintenance also affect the culture's growth and biomass. Photosynthetic microorganisms have evolved a multitude of regulatory mechanisms to maintain optimal photosynthetic efficiency and safeguard photosystems against the potential hazards of over-excitation and over-reduction. These adaptations enable their survival in a highly dynamic and intricate ecological setting.

2.5 Acclimation and Adaptation Strategies of Photosystems to Adverse Conditions

Photosynthetic organisms have developed a variety of mechanisms to adjust their photosynthetic activity in response to the environmental conditions, so as to prevent over-reduction and the generation of ROS. These regulatory mechanisms can respond to both short- and long-term dynamics of environmental variables (Eberhard et al. 2008; Walters 2005). The long-term acclimation response refers to the process of modifying the composition of the photosynthetic apparatus in order to adapt to changes in irradiance intensity. This adaptation involves regulating the production of chlorophyll and carotenoids. On a shorter temporal scale, photosynthetic organisms engage in regulatory mechanisms that modulate the efficiency of light absorption, notably through NPQ. This process facilitates the dissipation of surplus excitation energy as heat. The xanthophyll cycle, which produces zeaxanthin from already-existing violaxanthin, is also activated by strong illumination, adding to the protection from ROS (Havaux et al. 2007). The productivity of photosynthetic microorganisms in terms of biomass is significantly influenced by several factors, including the average irradiance per cell, mixing, gas exchange (specifically supply of CO_2 and removal of O_2), nutrition, and temperatures. When compared to land plants, microalgae show great metabolic adaptability, which allows them to change the biomass composition and metabolic flux in response to environmental stresses that affect photosynthesis. For instance, *Chlorella ohadii* is known to withstand rapid fluctuations in temperature and light intensities demonstrating the high degree of adaptability of microalgae to changing environmental factors. Molecular and biochemical analysis of this unique microalgal species revealed several interesting elements in photoprotection. In general, LHCSR protein is an essential part of the LHCII antenna complex and plays an important role in photoprotection in algae and plants. This particular protein is missing in *C. ohadii* due to the deletion of the corresponding genes and as a result the antenna size is significantly smaller than that of other genus members. Growth at high light levels, absence of LHCSR, absence of the zeaxanthin-violaxanthin, absence of NPQ cycle, and lack of a light response in carotenoid expression, all the elements contribute for a novel method of photoprotection in *C. ohadii*.

The potential explanation for *C. ohadii's* remarkable resilience may lie in the augmentation of PSII-cyclic electron flow (PSII-CEF), a process that involves the recycling of electrons from the acceptor side of PSII back to the donor side. The use of this mechanism for electron recycling presents a potential solution for energy production that circumvents the limitation of water scarcity. Previous studies have shown that this pathway plays a role in mitigating the over-excitation of algal PSIIs although with species-specific variations. By altering the redox potentials of these electron transfer cofactors near pheophytin and Q_A in cyanobacteria, known alterations in the amino acid sequence of the D1 subunit of PSII lead to increased fitness in response to light flow. Eukaryotic phototrophs only have one D1 form; hence, they are unable to use the process used by prokaryotic phototrophs to adjust to changing light intensity. Antenna complexes, formed by the arrangement of these pigments, are used to collect light energy and send it to the center of the reaction.

2.6 Conclusions

The understanding of structural, functional, and regulatory aspects of photosynthesis in microbial organisms gives an outlook on the functional dynamics of photosynthetic machinery in response to varying environmental conditions and light intensities. The chapter gives a comprehensive overview of the processes and structural aspects of photosynthesis in photosynthetic microorganisms, with a focus on the functional dynamics of photosystem complexes. Cyanobacteria and microalgae have structural differences in the arrangement of thylakoids and photosystem complexes compared to higher plants, affecting their light absorption and energy transfer. The two photosystem complexes—Photosystem II (PSII) and Photosystem I (PSI) absorb light at different wavelengths and play essential roles in electron transfer and ATP/NADPH generation. PSII is responsible for splitting water molecules and releasing oxygen as a by-product through the oxygen-evolving complex. Electrons flow through a series of redox reactions involving multiple electron carriers, leading to the generation of a proton gradient across the thylakoid membrane. The efficiency of photosynthetic machinery is regulated by factors such as light intensity, pigmentation, and pH gradients. Further reading of biochemical compositions and regulatory elements of photosystem in non-model photosynthetic microbes will give a new horizon in understanding the photosynthesis and its practical implications.

References

Allen JF (2002) Photosynthesis of ATP—electrons, proton pumps, rotors, and poise. Cell 110:273–276

Amunts A et al (2010) Structure determination and improved model of plant photosystem I. J Biol Chem 285:3478–3486

Anderson JM, Chow W (2002) Structural and functional dynamics of plant photosystem II. Philos Trans R Soc Lond B Biol Sci 357:1421–1430

Anderson LB et al (2002) Posttranslational modifications in the CP43 subunit of photosystem II. Proc Natl Acad Sci 99:14676–14681

Apostolova EL et al (2006) Relationship between the organization of the PSII supercomplex and the functions of the photosynthetic apparatus. J Photochem Photobiol B Biol 83:114–122

Arsalane W et al (1994) Influence of the pool size of the xanthophyll cycle on the effects of light stress in a diatom: competition between photoprotection and photoinhibition. Photochem Photobiol 60:237–243

Bhatti AF et al (2021) State transitions and photosystems spatially resolved in individual cells of the cyanobacterium Synechococcus elongatus. Plant Physiol 186:569–580

Blankenship RE (2021) Molecular mechanisms of photosynthesis. Wiley, Hoboken

Boyer PD (2002) A research journey with ATP synthase. J Biol Chem 277:39045–39061

Dau H, Haumann M (2007) Eight steps preceding O–O bond formation in oxygenic photosynthesis—a basic reaction cycle of the photosystem II manganese complex. Biochimica et Biophysica Acta (BBA)-Bioenergetics 1767(6):472–483

Eberhard S et al (2008) The dynamics of photosynthesis. Annu Rev Genet 42:463–515

Falkowski PG, Raven JA (2013) Aquatic photosynthesis. Princeton University Press, Princeton

Faseela P et al (2020) Chlorophyll a fluorescence parameter as indicators of a particular abiotic stress in rice. Photosynthetica 58:293–300
Gokhale Z, Sayre RT (2009) Photosystem II, a structural perspective. In: The chlamydomonas sourcebook. Elsevier, Amsterdam, pp 573–602
Havaux M et al (2007) Zeaxanthin has enhanced antioxidant capacity with respect to all other xanthophylls in Arabidopsis leaves and functions independent of binding to PSII antennae. Plant Physiol 145(4):1506–1520
Hill R, Bendall F (1960) Function of the two cytochrome components in chloroplasts: a working hypothesis. Nature 186:136–137
Jordan P et al (2001) Three-dimensional structure of cyanobacterial photosystem I at 2.5 Å resolution. Nature 411:909–917
Koua FHM et al (2013) Structure of Sr-substituted photosystem II at 2.1 Å resolution and its implications in the mechanism of water oxidation. Proc Natl Acad Sci 110(10):3889–3894
Krewald V et al (2015) Metal oxidation states in biological water splitting. Chem Sci 6(3):1676–1695
Kumar S et al (2017) Molecular insights into photosynthesis and carbohydrate metabolism in Jatropha curcas grown under elevated CO_2 using transcriptome sequencing and assembly. Sci Rep 7(1):11066
Lavaud J et al (2004) General features of photoprotection by energy dissipation in planktonic diatoms (Bacillariophyceae) 1. J Phycol 40:130–137
Lubitz W et al (2019) Water oxidation in photosystem II. Photosynth Res 142(1):105–125
Madhana Sekhar K et al (2014) Persistent stimulation of photosynthesis in short rotation coppice mulberry under elevated CO_2 atmosphere. J Photochem Photobiol B 137:21–30. https://doi.org/10.1016/j.jphotobiol.2014.05.001
Mamedov M et al (2015) Primary electron transfer processes in photosynthetic reaction centers from oxygenic organisms. Photosynth Res 125:51–63
Melis A (2009) Solar energy conversion efficiencies in photosynthesis: minimizing the chlorophyll antennae to maximize efficiency. Plant Sci 177(4):272–280
Murata N et al (2007) Photoinhibition of photosystem II under environmental stress. Biochim Biophys Acta Bioenerg 1767:414–421
Mustardy L et al (2008) The three-dimensional network of the thylakoid membranes in plants: quasihelical model of the granum-stroma assembly. Plant Cell 20:2552–2557
Nelson N, Junge W (2015) Structure and energy transfer in photosystems of oxygenic photosynthesis. Annu Rev Biochem 84:659–683
Nickelsen J, Rengstl B (2013) Photosystem II assembly: from cyanobacteria to plants. Annu Rev Plant Biol 64:609–635
Ort DR, Baker NR (2002) A photoprotective role for O_2 as an alternative electron sink in photosynthesis? Curr Opin Plant Biol 5:193–198
Pérez-Navarro M et al (2016) Recent developments in biological water oxidation. Curr Opin Chem Biol 31:113–119
Puthiyaveetil S et al (2012) Oxidation–reduction signalling components in regulatory pathways of state transitions and photosystem stoichiometry adjustment in chloroplasts. Plant Cell Environ 35:347–359
Ranadheer P et al (2019) Non-lethal nitrate supplementation enhances photosystem II efficiency in mixotrophic microalgae towards the synthesis of proteins and lipids. Bioresour Technol 283:373–377
Reddy KS et al (2019) Hydraulic dynamics and photosynthetic performance facilitate rapid screening of field grown mulberry (Morus spp.) genotypes for drought tolerance. Environ Exp Bot 157:320–330
Sawyer A, Winkler M (2017) Evolution of Chlamydomonas reinhardtii ferredoxins and their interactions with [FeFe]-hydrogenases. Photosynth Res 134:307–316

Sekhar KM et al (2015) Differential responses in photosynthesis, growth and biomass yields in two mulberry genotypes grown under elevated CO_2 atmosphere. J Photochem Photobiol B 151:172–179. https://doi.org/10.1016/j.jphotobiol.2015.08.008

Sharma A et al (2020) Photosynthetic response of plants under different abiotic stresses: a review. J Plant Growth Regul 39:509–531

Shevela D et al (2021) Photosystem II. eLS 2:1–16

Shikanai T (2007) Cyclic electron transport around photosystem I: genetic approaches. Annu Rev Plant Biol 58:199–217

Sreeharsha RV, Reddy AR (2015) Dynamics of vegetative and reproductive growth patterns in Pigeonpea (*Cajanus cajan* L.) grown under elevated CO_2. Procedia Environ Sci 29:147–148

Sreeharsha RV, Venkata Mohan S (2021) Symbiotic integration of bioprocesses to design a self-sustainable life supporting ecosystem in a circular economy framework. Bioresour Technol 326:124712

Sreeharsha RV et al (2019) Mitigation of drought-induced oxidative damage by enhanced carbon assimilation and an efficient antioxidative metabolism under high CO_2 environment in pigeonpea (*Cajanus cajan* L.). Photosynth Res 139(1):425–439

Sreeharsha RV et al (2023) Orienting biodiesel production towards sustainability and circularity by tailoring the feedstocks and processes. J Clean Prod 137526

Szabó I et al (2005) Light and oxygenic photosynthesis: energy dissipation as a protection mechanism against photo-oxidation. EMBO Rep 6:629–634

Umena Y et al (2011) Crystal structure of oxygen-evolving photosystem II at a resolution of 1.9 Å. Nature 473:55–60

van Amerongen H, Croce R (2013) Light harvesting in photosystem II. Photosynth Res 116:251–263

Walters RG (2005) Towards an understanding of photosynthetic acclimation. J Exp Bot 56:435–447

Yano J, Yachandra V (2014) Mn4Ca cluster in photosynthesis: where and how water is oxidized to dioxygen. Chem Rev 114(8):4175–4205

Zehr JP, Kudela RM (2009) Photosynthesis in the open ocean. Science 326:945–946

Chapter 3
Regulatory Mechanisms of Photosynthetic CO_2 Fixation in Unicellular Microbes

Abstract Aquatic photoautotrophs, such as microalgae and cyanobacteria, contribute 50% of carbon assimilated into organic matter through their carbon fixation ability. The Calvin-Benson cycle, a complex metabolic pathway, reduce carbon dioxide (CO_2) and drive glucose production. Efficiency of carbon sequestration depends on the uninterrupted supply of reducing equivalents from light reactions and RuBp regeneration capacity. Regulation of the Calvin cycle in photosynthetic microbes involves glyceraldehyde-3-phosphate dehydrogenase, phosphoribulokinase, and CP12 protein complex. Rubisco produces two 3-phosphoglycerate molecules when CO_2 is fixed to the acceptor molecule, ribulose-1,5-bisphosphate (RuBP). The enzyme has competitive carboxylase and oxygenase activities that rely on the balance of oxygen and CO_2 at its active site. The RUBISCO enzyme has evolved in various forms with varying kinetic characteristics in autotrophic organisms. The CO_2 concentrating mechanism (CCM) is an efficient approach to carbon acquisition used by microalgae and other photosynthetic microbes to survive and propagate when CO_2 concentration limits photosynthesis. Photorespiration is a crucial metabolic process that helps to cope with high oxygen levels and prevent damage to photosynthetic machinery. Iron-sulfur clusters, malate shuttles, and light-enhanced respiration (LER) play essential roles in photorespiration and photosynthesis. Various approaches have been employed to improve algal growth and productivity cost-efficiently. Genetic engineering tools have been utilized to reduce antenna size, improve photosynthetic efficiency, and increase carbon fixation. Overexpression of RuBisCO activase, sedoheptulose bisphosphatase, and fructose bisphosphate, aldolase has also been shown to improve growth and photosynthesis. Recent developments in metagenomics-based discoveries, CRISPR-based genome editing, and adaptive evolution methods could facilitate the development of efficient microalgal strains for CO_2 fixation.

Keywords Glyceraldehyde-3-phosphate dehydrogenase · Phosphoribulokinase · RuBisCO activase · Sedoheptulose bisphosphatase · RUBISCO · Calvin cycle

R. V. Sreeharsha, S. Venkata Mohan, *Microbial Photosynthesis*,
https://doi.org/10.1007/978-981-97-1253-3_3

3.1 Introduction

Oxygenic photosynthesis is one of the major sinks for atmospheric carbon thus producing organic substances form inorganic carbon. The reducing equivalents and energy compounds produced during the light reaction are used in the Calvin-Benson cycle wherein CO_2 is reduced to drive the production of glucose. It is the most complex and well-conserved metabolic pathway among living organisms that takes place in chloroplasts and involves a series of enzymatic chemical reactions. RUBISCO which is the crucial and major enzyme in carbon fixation is also the most abundant enzyme on earth biome. In the current scenario, the carbon fixation pathway is significant in the context of continuously rising CO_2 levels (Sreeharsha et al. 2019; Sreeharsha and Reddy 2015). Interestingly, the majority of carbon assimilated (50%) into organic matter is contributed by aquatic photoautotrophs, among which microalgae and cyanobacteria play a significant role (Vecchi et al. 2020). Compared to the best terrestrial systems, algae have been estimated to produce 2 to 10 times more biomass per land surface unit (Sayre 2010). Their ability to reduce photorespiration by carbon concentrating mechanism enhances photosynthesis leading to increased biomass.

The atmospheric CO_2 levels have increased from 280 μmol mol^{-1} since the pre-industrial era to the current 410 μmol mol^{-1}. Subsequently, the CO_2 levels are responsible for increasing the temperature and this leads to the greenhouse gas effect. The elevated CO_2 concentrations are having adverse effects on natural ecosystems and showing negative impact on agricultural productivity worldwide (Sreeharsha et al. 2015; Unnikrishnan et al. 2021). Photosynthetic organisms show varied responses to increased CO_2 concentrations wherein the maximal photosynthetic efficiency showed an increase initially and thereby reduced due to feedback inhibition from sink tissues (Kumar et al. 2017; Sekhar et al. 2015). The efficiency of carbon sequestration depends on the uninterrupted supply of reducing equivalents from light reactions as well as RuBp regeneration capacity to feed the Calvin cycle. In higher eukaryotes, the efficient development of sink tissues including above and below ground biomass will determine the source carbon uptake. Also, the biochemical regulation of stomatal opening controls the intracellular CO_2 concentrations that in turn influence the RUBISCO kinetics (Madhana Sekhar et al. 2014). However, in the case of photosynthetic microbes, the regulation of CO_2 fixation is dependent on various other abiotic factors such as nutrients, pH, temperature and growth stage of the organism (Ranadheer et al. 2019). Further, photosynthetic enzymes which are under tight regulation also play a crucial role in carbon fixation. The present chapter gives a detailed overview of the Calvin cycle and its molecular regulation in microalgae and cyanobacteria. It also presents the kinematics of photorespiration and prospects to enhance carbon fixation in photosynthetic microbes.

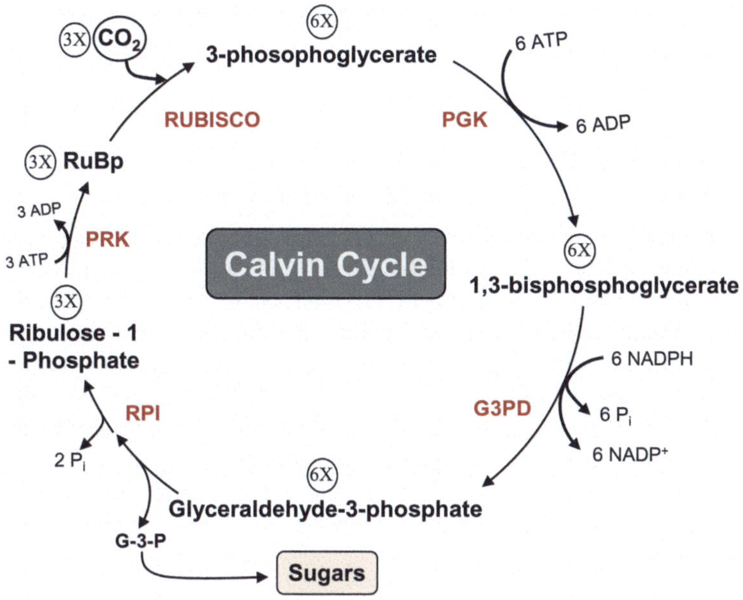

Fig. 3.1 Metabolic steps and corresponding enzymes involved in Calvin cycle

3.2 Calvin-Benson-Bassham (CBB) Cycle

Calvin cycle is a redox process involving the conversion of CO_2 into sugars in three phases—carboxylation, reduction, and regeneration (Fig. 3.1) (Curien et al. 2016). During *Carbon Fixation* phase, CO_2 from the atmosphere is captured by a five-carbon sugar molecule called ribulose-1,5-bisphosphate (RuBP). The enzyme involved in this step is ribulose-1,5-bisphosphate carboxylase/oxygenase, or RuBisCO. It catalyzes the reaction between CO_2 and RuBP, forming an unstable six-carbon compound that immediately splits into two three-carbon molecules called 3-phosphoglycerate (3-PGA). In *Reduction Phase,* ATP and NADPH generated during the light-dependent reactions of photosynthesis are used to convert the 3-PGA molecules into a different three-carbon compound called glyceraldehyde-3-phosphate (G3P). For every three molecules of CO_2 that enter the cycle, six molecules of G3P are produced. One of these molecules will eventually become glucose, while the others continue through the cycle. *Regeneration of RuBP* sustains the cycle, by diverting G3P molecules to regenerate RuBP through a series of enzymatic reactions with the utilization of ATP. This step ensures that RuBP is continually available for carbon fixation. Finally, in *Glucose Formation* phase, one of the G3P molecules produced in the reduction phase is used to create glucose and other carbohydrates through a series of enzymatic reactions outside the Calvin cycle. Glucose and other carbohydrates serve as energy storage molecules and are essential for growth and survival of the organism. The remaining G3P molecules not used for RuBP regeneration or glucose formation are used for various metabolic processes such as the synthesis of amino acids, lipids, and other organic compounds.

3.3 Regulation of Calvin Cycle

Multiple enzymes are involved in the Calvin cycle to regulate the metabolic flux and product formation. There is considerable diversity exist in the structure and amino acid composition of Calvin cycle enzymes of microalgae and cyanobacterial species making them unique regulatory elements (Table 3.1; Fig. 3.2). One of the regulatory mechanisms that controls the carbon fixation in photosynthetic microbes involves glyceraldehyde-3-phosphate dehydrogenase (GAPDH) and phosphoribulokinase (PRK), and CP12 protein complex. Complex formation often occurs under the regulation of thioredoxins as well as pyridine nucleotides. Under light, this complex dissociates to the reactivation of GAPDH and PRK (Gurrieri et al. 2021). The organic acids generated by RUBISCO are utilized by GAPDH to help in the reduction phase of sugars.

3.3.1 *PRK-GAPDH-CP12 Mediated Regulation*

PRK regenerates the Rubisco substrate with a single ATP. PRK is classified into three types: plant, archaeal, and bacterial. Plant PRK is found largely in plants, most cyanobacteria, and photosynthetic protists. Archaeal and bacterial PRK is found in proteobacteria and oceanic cyanobacteria. PRK belongs to the family of nucleoside/nucleotide kinase superfamily. Plant and archaeal PRK are dimeric, while bacterial PRK is octameric with monomers. There exist four distinct forms of GAPDH genes, namely GAPA, GAPB, GAPC, and GAPCp. Subsequently, these isoforms combine

Table 3.1 structural features of different enzymes involved in Calvin cycle in cyanobacteria and microalgae

	Number of amino acids		Molecular weight (KDa)	
Enzymes	Cyanobacteria	Microalgae	Cyanobacteria	Microalgae
Rubisco 1,5-bisphosphate carboxylase	460	412	50.6	45.32
Phosphoglycerate kinase	399	400	43.89	44
Glyceraldehyde-3, phosphate dehydrogenase	339	313	37.29	34.43
Ribose phosphate isomerase	509	315	55.99	34.65
Phosphoribulokinase	102	102	11.22	11.22
Transketolase	73.81	163.79	671	1489
Fructrose-1,6 bisphosphatase	36.74	255.97	334	2327
Sedoheptulose-7, bisphosphate	38.6	42.79	346	389

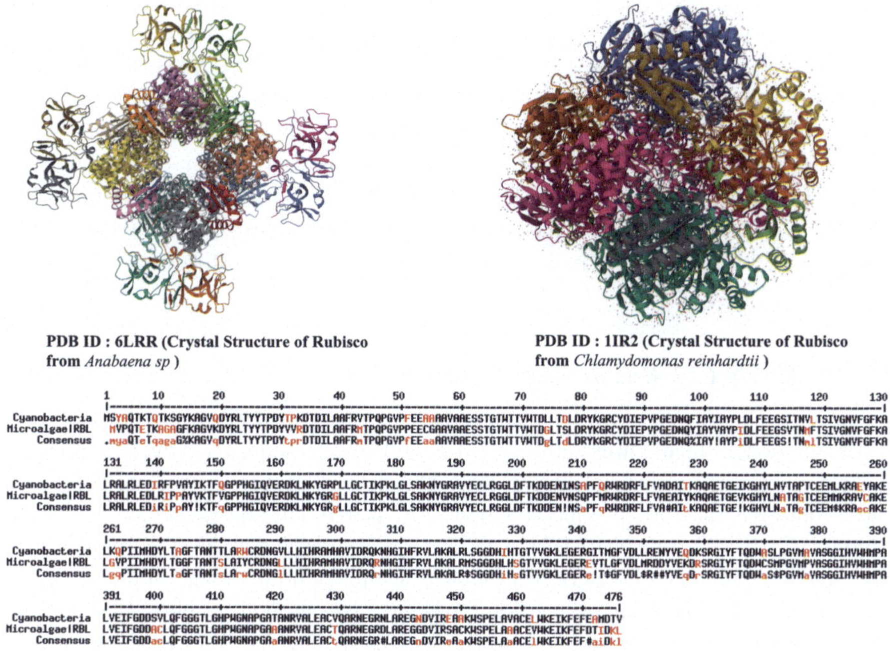

Fig. 3.2 Structural features of RUBISCO enzyme in cyanobacteria and microalgae. Represented the RUBISCO structure obtained from the Pfam database and the sequence similarity in amino acid sequence of both the cyanobacteria and microalgae

to generate proteins consisting of four tetrameric structures, including A4 and A2B2 within the chloroplast, C4 in the cytoplasm, and CP4 in plastids. The A4 and C4-GAPDH isoforms only exist in green algae and cyanobacteria. CP12 is an intrinsically disordered protein of about 80 amino acids, however, upon interaction with GAPDH and PRK, it undergoes a conformational change and acquires a specific three-dimensional structure. The first step for assembling the ternary complex with PRK requires the association of GAPDH-CP12. The prerequisite for this association is GAPDH loaded with NAD(H) rather than NAPDH. Then these complexes form a ternary complex with PRK. The ternary complex is composed of two tetramers of glyceraldehyde-3-phosphate dehydrogenase (GAPDH) and two dimers of phosphoribulokinase (PRK), which are connected by four oxidized chloroplast protein 12 (CP12) molecules. The CP12 complex plays a crucial role in the regulation of the Calvin-Benson cycle in cyanobacteria during alternating light and dark conditions. In plants, enzymes involved in the Calvin cycle, like GAPDH, FBPase, SBPase, and PRK, are activated by the reduction of intramolecular disulfide bridges via ferredoxin/thioredoxin.

On the other hand, based on comparative analysis, a lack of redox-sensitive cys residues necessary for light/dark regulation was observed in the Calvin cycle enzymes of algae and cyanobacteria (Tamoi and Shigeoka 2015). A new type of FBP/SBPase was observed in cyanobacterial cells that hydrolyzes both fructose 1,6-bisphosphate and sedoheptulose 1,7-bisphosphate, with different primary structures from typical FBPase and SBPase. This confirms that cyanobacterial and algal enzymes are not regulated via ferredoxin/thioredoxin redox regulation under light/dark conditions. It is regulated by the association/dissociation of PRK/CP12/GAPDH complex based on the in vivo NADP(H)/NADH ratio. Enzymes like FBP/SBPase, CP12, GAPDH, and PRK are targeted to improve algal growth and photosynthesis. Rubisco is another protein targeted in engineering to improve algal growth.

3.3.2 *RuBisCO-Based Regulation*

Most cyanobacteria and microalgae rely on the Rubisco and Calvin cycle to convert inorganic carbon to organic matter. With the help of Rubisco, two molecules of 3-phosphoglycerate are produced when CO_2 is fixed to the acceptor molecule ribulose-1,5-bisphosphate (RuBP). Additionally, Rubisco catalyzes the oxygenation of RuBP, forming one phosphoglycerate and one phosphoglycolate molecule rather than two phosphoglycerate molecules. One of the two carbons in phosphoglycolate can be retrieved and converted to phosphoglycerate by the process of the Photosynthetic Carbon Oxidation Cycle (PCOC) during photorespiration, but the other carbon is lost as CO_2, demonstrating the inefficiency of the carbon assimilatory process. Rubisco has competitive carboxylase and oxygenase activities that rely on the balance of oxygen and CO_2 at the enzyme's active site. The structural diversity of Rubisco among the phylogenetic groups makes this enzyme unique in its properties in each of the species (Fig. 3.2). Rubisco has evolved in various forms with various kinetic characteristics in autotrophic organisms. In addition to Forms I, II, and III, which are the three primary forms of Rubisco that are known, there is a fourth category (Form IV) of Rubisco-like proteins that lack carboxylase activity but may be involved in S metabolism (Tabita et al. 2008). Most microalgae and cyanobacteria have variants of Rubisco that are classified as Form I and include eight large and eight small subunits (L_8S_8). The large subunit is ~55 kDa, and the small subunit is ~15 kDa. The enzyme's active center is located in the large subunit, and the small subunit acts for stability and catalytic efficiency. The variants of Form I are based on the difference in the primary sequence of the large subunit (Kupriyanova et al. 2023). Form IA Rubisco, produced by all marine Prochlorococcus species and many marine *Synechococcus* species, was gained through lateral gene transfer from an autotrophic proteobacterium (Raven and Hurd 2012). However, the majority of freshwater and marine cyanobacteria possess the ancestral Form IB Rubisco, which was passed to chlorophyte and euglenophyte algae through secondary endosymbiosis and to glaucocystophyte and chlorophyte algae through primary endosymbiosis. While Form II Rubisco only has two large subunits (L2), Form II Rubisco was acquired by the ancient dinoflagellates and Chromera velia1, both alveolates, by lateral gene transfer from an autotrophic proteobacterium. Form III

Rubiscos have more complex structures as they lack small subunits. Different forms of Rubisco have different kinetic properties. Cyanobacteria and microalgae cannot saturate the current atmospheric CO_2 concentration at Rubisco active centers. CO_2 has a slight advantage over O_2 as a substrate due to the low affinity for CO_2 and the low CO_2/O_2 specificity of cyanobacterial Rubisco. However, Rubisco in green microalgae has an affinity for CO_2 compared to cyanobacteria. To overcome this problem, cyanobacteria and microalgae employ CCM as an add-on to the RPP cycle.

3.3.3 CO_2 Concentrating Mechanisms in Microbes

CO_2 concentrating mechanism (CCM) is an efficient approach of carbon acquisition used by microalgae and other photosynthetic microbes to survive and propagate when CO_2 concentration limits photosynthesis. It is an adaptive strategy by cyanobacteria and microalgae under unfavorable conditions of limited CO_2 and increased O_2. It is an integral part of algal photosynthesis, growth, and metabolism. The ability to counterpoise the inefficiency of Rubisco is the central aspect of CCM to improve photosynthetic efficiency (Wang et al. 2015). The primary mechanism of CCM is as follows—the first step involves the creation of an HCO_3^- pool higher in Ci concentration compared to the environment, and in the second step, this HCO_3^- is used to generate CO_2 molecules near Rubisco active centers. These are single-cell-based CO_2 enrichment mechanisms depending on the multiple Ci uptake systems. The most effective Ci uptake system leads to a 1000-fold increase in intracellular Ci in the form of HCO_3^-. Microalgal photosynthesis is set apart from C4 photosynthesis, which has active CO_2 and HCO_3^- uptake systems. This is an essential survival feature as most organisms are aquatic. Since CO_2 diffusion is slower in water, HCO_3^- is a more dominant form. Hence, both types of uptake systems depend on the abundance of each source to switch the different modes of Ci acquisition based on the environment. However, Rubisco can only utilize CO_2, so carbonic anhydrase at the site of Rubisco converts HCO_3^- to CO_2. Rubisco is highly localized in the pyrenoids microcompartment, where the CO_2 concentration can be increased significantly. They are known as carboxysomes in cyanobacteria and pyrenoids in eukaryotic algae. Also, there is a CO_2 barrier to prevent the leakage of CO_2. While most C4 photosynthesis is expressed constitutively, CCM is an inducible mechanism. Ci uptake happens at the plasma membrane, chloroplast envelope, and chloroplasts. Transcriptome studies have identified several Ci uptake components including Ci transporters such as HLA3, LCI1, LCIA, CCP1, CCP2, and RHPs. Similarly, HLA3 and LCIA are HCO_3^- transporters at the plasma membrane activated at deficient CO_2 acclimation states, while LCIB, LCIC, LCID, and LCIE are involved in CO_2 uptake. Furthermore, carbonic anhydrase, which is involved in enriching CO_2 levels in the chloroplasts, is considered essential for CCM. Carbonic anhydrases are a group of proteins wherein CAH 1 and 2 are located in the periplasm and CAH 4 and 5 are located in the mitochondria. While CAH 3 is found in the thylakoid lumen and CAH 6 in the chloroplast stroma (Wang et al.

2015). Various approaches are under investigation to engineer Calvin enzymes, Rubisco, and CCM to improve photosynthesis in algae and are discussed in Chap. 8.

3.4 Photorespiration

Photorespiration is a complex metabolic process that occurs in algae, allowing them to cope with high oxygen levels and prevent damage to photosynthetic machinery (Fig. 3.3). Understanding the regulation of photorespiration is crucial for improving the efficiency of photosynthesis and enhancing biomass production. During photorespiration oxygen competes with CO_2 as a substrate for the enzyme Rubisco leading to the production of toxic metabolites and results in a loss of energy and carbon fixation efficiency. Understanding photorespiration in microalgae is important for improving their growth and targeted product generation. The thorough characterization of the rate of photorespiration in the presence of light is an essential ecophysiological concept that has yet to be fully understood. It has been noticed that

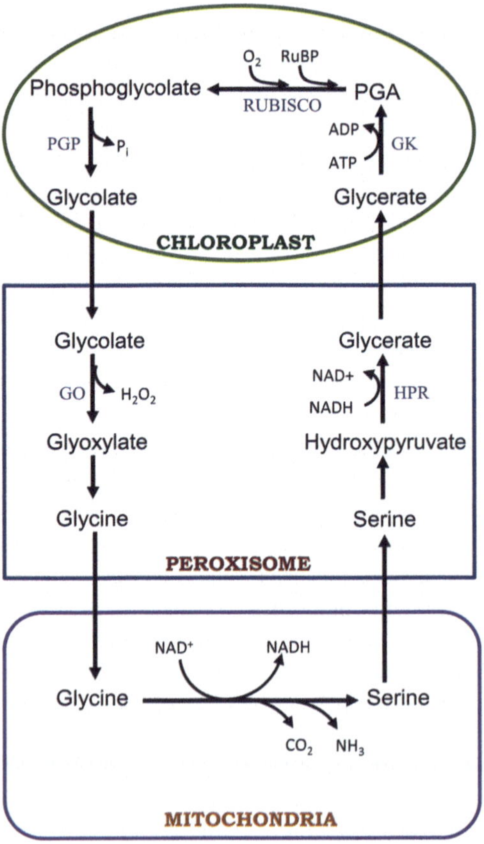

Fig. 3.3 Schematic representation of photorespiration, cellular events, and key enzymes involved in the process

decarboxylation, which is a crucial process in photorespiration, contributes around 22% to the overall net absorption of CO_2 when the tricarboxylic acid cycle is fully operational. The provided estimate was substantiated through the computation of decarboxylation rates necessary for the synthesis of prominent biomass constituents at commonly observed levels.

Iron-sulfur clusters that are assembled in mitochondria, plastids, and cytosol play essential roles as cofactors in many enzymes involved in photorespiration and photosynthesis. The physiological functions of malate shuttle involve transportation of malate across membranes which play a crucial role in balancing the redox poise between different cellular compartments. They have been implicated in photorespiration and carbon concentrating mechanisms in photosynthetic organisms. Light-enhanced respiration (LER) wherein increase in respiratory oxygen uptake takes place after illumination is linked to the increase in respiratory substrates by photosynthetic CO_2 assimilation. LER can contribute to energy production and metabolic homeostasis under light conditions. Mixotrophic metabolism allows diatoms to grow on both light and organic compounds, providing them with trophic flexibility. Non-steroidal anti-inflammatory drugs (NSAIDs) exerted inhibitory effects on *Scenedesmus obliquus* growth, by affecting cellular ultrastructure, chlorophyll content, and chlorophyll fluorescence parameters such as F0, Fm, Fv/Fm, Y(II), ETR, and qP (Wang et al. 2020). Similarly, the growth of coastal diatom *Thalassiosira pseudonana* under iron (Fe) limitation up-regulates genes involved in intracellular protein recycling, thereby decreasing dependence on extracellular nitrogen, carbon, and Fe with a concurrent increase in photorespiration and pentose phosphate pathway proteins (Nunn et al. 2013). Research conducted on the cyanobacterium *Synechocystis* has provided insights into the regulatory role of the transcription factor RpaB in several biological processes. These processes include photoprotection, cyclic electron flow, state transitions, photorespiration, and interplay with other transcription factors like NtcA or FurA (Riediger et al. 2019).

3.5 Approaches to Improve Carbon Sequestration

Algae and cyanobacteria are considered an important commercial source for biofuels, feed, and other high-value products. Hence various methods have always been utilized to improve their growth and productivity cost-efficiently. However, engineering photosynthetic pathways for improving algal productivity is not well established. Energy absorbed by photosystems is lost by non-photochemical quenching (NPQ). So, genetic engineering oriented reducing antenna size, thereby reducing NPQ and improving photosynthetic efficiency (Naduthodi et al. 2021). Chloroplast signal recognition particle protein (CpSRP) has been targeted for reducing antenna size. CpSRP translocates light-harvesting proteins to the thylakoid membrane. Improved photosynthetic efficiency and biomass productivity were observed in some microalgal species. However, some were adversely affected as part of antenna truncation (Shin et al. 2016). Studies showing the replacement of DI subunit in PSII of *Chlamydomonas* from cyanobacterium *Synechococcus elongatus* showed improved PSII photochemistry in some variants under particular conditions. However, studies showing the replacement of PSII by plant PSII did not show much improvement in photosynthesis. This might be because replacing a single protein in the whole photosystem might not be sufficient, as other factors, like interactions between multiple proteins, need to be considered. Extensive research and optimization need to be done in this aspect of engineering. To improve the Rubisco activity in the microalgal system, overexpression of RuBisCO activase has been used to increase the levels of active Rubisco, showing improved growth rate (32%) and biomass accumulation (46%) in *Nannochloropsis oceanica.* Replacing native rubisco variants with modified rubisco variants in a natural environment improves growth, but under high CO_2 conditions, like in a photobioreactor may not work.

In addition to Rubisco, additional enzymes from the Calvin cycle, such as Sedoheptulose bisphosphatase (SBP) and fructose bisphosphate aldolase (FBA), are also considered as they are rate-limiting steps. Overexpression of Calvin enzymes has increased growth and photosynthesis in cyanobacteria like *Synechocystis, Synechococcus,* and *Anabaena* spp. Cyanobacterial SBP and FBA overexpression in the microalgae *Chlorella vulgaris* and *Euglena gracilis* showed improved biomass production by twofold. Engineering improved variants of the Calvin enzymes with the help of directed mutation and genome editing, alongside improvements to RuBisCO, can increase CO_2 fixation via the CBB cycle. Microalgae employ CCM to reduce photorespiration. Many enzymes and genes are involved in the CCM pathway. Engineering these CCM pathway products can also improve photosynthesis and reduce photorespiration. The bicarbonate transporter from *C. reinhardtii* was overexpressed in *N. salina,* showing enhanced growth rate and biomass accumulation.

A significant challenge in growing algae commercially is the high cost of growth conditions like phototrophic conditions. Hence research on efficient ways to improve photosynthesis by increasing CO_2 fixation is employed to increase algal growth without the high-end cost. The development of natural or artificial CO_2 fixation routes that would result in microalgal hosts with increased productivity could be facilitated by recent developments in metagenomics-based discoveries, CRISPR-based genome editing, and adaptive evolution methods.

3.6 Conclusion

The Calvin cycle and the carbon concentrating mechanisms are crucial in improving the algal growth and productivity. The dark reaction of photosynthesis or Calvin cycle is divided into carbon fixation, reduction, and regeneration phases which are catalyzed by crucial rate-limiting enzymes. Understanding and studying these enzymes is essential to modify them genetically so as to enhance the efficiency of the organisms. Photorespiration is another important cellular activity in microalgae which is futile and produces no energy equivalents. The partial pressure of CO_2 and

O_2 and binding affinity for RUBISCO determines the fate of the cellular event. There are several studies showing the significance of reactive oxygen species that are generated during the photorespiration process in cellular signaling in plants. However, the practical implications of photorespiration in photosynthetic microbes, particularly, in biorefinery applications needs to be studied further. Also, active research is going on to modulate the expression of Calvin cycle enzymes to enhance the growth and productivity of photosynthetic microbes.

References

Curien G et al (2016) The water to water cycles in microalgae. Plant Cell Physiol 57:1354–1363

Gurrieri L et al (2021) Calvin–Benson cycle regulation is getting complex. Trends Plant Sci 26:898–912

Kumar S et al (2017) Molecular insights into photosynthesis and carbohydrate metabolism in Jatropha curcas grown under elevated CO_2 using transcriptome sequencing and assembly. Sci Rep 7:11066. https://doi.org/10.1038/s41598-017-11312-y

Kupriyanova EV et al (2023) Adapting from low to high: an update to CO_2-concentrating mechanisms of cyanobacteria and microalgae. Plan Theory 12:1569

Madhana Sekhar K, Rachapudi VS, Mudalkar S, Reddy AR (2014) Persistent stimulation of photosynthesis in short rotation coppice mulberry under elevated CO_2 atmosphere. J Photochem Photobiol B 137:21–30

Naduthodi MIS et al (2021) Synthetic biology approaches to enhance microalgal productivity. Trends Biotechnol 39:1019–1036

Nunn BL et al (2013) Diatom proteomics reveals unique acclimation strategies to mitigate Fe limitation. PloS One 8:e75653

Ranadheer P et al (2019) Non-lethal nitrate supplementation enhances photosystem II efficiency in mixotrophic microalgae towards the synthesis of proteins and lipids. Bioresour Technol 283:373–377

Raven JA, Hurd CL (2012) Ecophysiology of photosynthesis in macroalgae. Photosynth Res 113:105–125

Riediger M et al (2019) Biocomputational analyses and experimental validation identify the regulon controlled by the redox-responsive transcription factor RpaB. iScience 15:316–331

Sayre R (2010) Microalgae: the potential for carbon capture. Bioscience 60:722–727

Sekhar KM, Sreeharsha RV, Reddy AR (2015) Differential responses in photosynthesis, growth and biomass yields in two mulberry genotypes grown under elevated CO_2 atmosphere. J Photochem Photobiol B 151:172–179

Shin W-S et al (2016) Truncated light-harvesting chlorophyll antenna size in Chlorella vulgaris improves biomass productivity. J Appl Phycol 28:3193–3202

Sreeharsha RV et al (2015) Delayed flowering is associated with lack of photosynthetic acclimation in Pigeon pea (Cajanus cajan L.) grown under elevated CO_2. Plant Sci 231:82–93

Sreeharsha RV, Reddy AR (2015) Dynamics of vegetative and reproductive growth patterns in Pigeonpea (Cajanus cajan L.) grown under elevated CO_2. Procedia Environ Sci 29:147–148

Sreeharsha RV et al (2019) Mitigation of drought-induced oxidative damage by enhanced carbon assimilation and an efficient antioxidative metabolism under high CO_2 environment in pigeonpea (Cajanus cajan L.). Photosynth Res 139(1):425–439

Tabita FR et al (2008) Distinct form I, II, III, and IV Rubisco proteins from the three kingdoms of life provide clues about Rubisco evolution and structure/function relationships. J Exp Bot 59:1515–1524

Tamoi M, Shigeoka S (2015) Diversity of regulatory mechanisms of photosynthetic carbon metabolism in plants and algae. Biosci Biotechnol Biochem 79:870–876

Unnikrishnan DK et al (2021) Growth, seed yield and nutritional characteristics of pigeonpea grown under elevated CO_2 atmosphere. Acta Physiol Plant 43:1–15

Vecchi V et al (2020) Potential and challenges of improving photosynthesis in algae. Plan Theory 9:67

Wang Y et al (2015) The CO_2 concentrating mechanism and photosynthetic carbon assimilation in limiting CO_2: how chlamydomonas works against the gradient. Plant J 82:429–448

Wang H et al (2020) Photosynthetic toxicity of non-steroidal anti-inflammatory drugs (NSAIDs) on green algae scenedesmus obliquus. Sci Total Environ 707:136176

Part II
Resource Valorization in a Biorefinery Approach

Chapter 4
Methods in Photosynthetic Physiology, Cultivation and Harvesting of Photosynthetic Microbes

Abstract The crucial aspects of cultivation of photosynthetic microbes are assessing photosynthetic physiology and optimizing bioreactors and biomass processing. Several instruments including Pulse Amplitude Modulator, Fluorometer, and Infrared Gas Analyzer measure photosynthetic dynamics. Fast kinetics of chlorophyll *a* fluorescence and light response curves give holistic understanding of photosystem complexes to different environmental conditions. OJIP fluorescence transients' analysis and corresponding K, L, and P bands investigate the kinetics, regulation, and functional significance of photosystems and other photosynthetic machinery. The dissipation of excess energy occurs through passive mechanisms such as heat and fluorescence emitted by light-harvesting complexes, which play a crucial role in the production of Y(NO) and Y(NPQ). Other than photosynthetic physiology, designing of bioreactors and downstream processing are crucial elements in microbial biorefinery, impacting the acquisition, alteration, extraction, and purification of high-value compouunds. Two-stage cultivation of microalgae optimizes growth and productivity for various applications thus reducing competition for resources. Algae and bacteria symbiosis is beneficial for nutrient exchange, gene transfer, and signal transduction and highly useful in fixed bed photobioreactors, airlift reactors, and photobioreactors. Photobioreactors are artificially constructed systems that facilitate light-dependent photosynthesis, allowing for higher algal yields. Airlift reactors have a high degree of flexibility and are well-suited for integration into biorefineries and integrated power-generating systems. Horizontal tubular bioreactors are space-efficient and helical tubular bioreactors are space-efficient but require larger pipes and tubes for scaling up. Flat plate bioreactors aim to maximize efficiency from sunlight but have drawbacks like disrupted flow control and increased oxygen content. High-rate algal ponds are sustainable, cost-effective, and have applications in cattle and aquaculture feed, soil conditioner, and biofuel production. The present chapter outlines the techniques involved in photosynthetic physiology, describes the instrumentation and methodology. Further, the chapter gives a detailed overview of the culturing and processing techniques of photosynthetic microbes.

R. V. Sreeharsha, S. Venkata Mohan, *Microbial Photosynthesis*,
https://doi.org/10.1007/978-981-97-1253-3_4

Keywords Cocultivation · Fluorometer · Photobioreactors · Biomass processing · OJIP transients · Chlorophyll a fluorescence

4.1 Introduction

Photosynthetic microorganisms harness sunlight and atmospheric CO_2 during photosynthesis process. The crucial events of photochemical quenching and carbon sequestration such as intracellular CO_2 concentrations, rate of water splitting and oxygen evolution, and rate of electron transport chain will determine the growth and biomass productivity of the organism (Sekhar et al. 2015; Kumar et al. 2017). It is worth noticing that photosystem II and photosystem I are crucial in electron transport chain and generation of reducing energy equivalents which are utilized in subsequent metabolic processes. The readers have studied the structural and functional aspects of PSII and PSI in earlier chapters. This chapter describes various techniques useful to study the functional dynamics of these photosystem complexes and instruments available to give readers a comprehensive understanding of the laboratory protocols when working with liquid photosynthetic cultures. The nomenclature and terminology of photosystem parameters are distinguished and varied in the literature. There are unique techniques available to measure PSII and PSI independently, which will provide exhaustive data on photochemical events to study the subsequent downstream processes. The effect of photosynthetic photon flux density (PPFD) and actinic light on ETC and photochemical yield will give finest apprehension of the photosystem dynamics to different environmental conditions.

Culturing microbes with optimum light, pH, temperature, and nutrients is an important milestone in achieving targeted biomass and products. The photosynthetic organisms can be cultivated under photoautotrophic, mixotrophic, and heterotrophic modes based on the availability of light and CO_2 as energy and carbon sources, respectively. Earlier studies on microalgae have demonstrated higher growth rates and biomass productivity under mixotrophic conditions as the organism is getting benefited from both natural carbon fixation and organic or inorganic carbon sources. Further, the cultivation can be done in open or closed bioreactors which are having their own advantages and limitation. There has been significant development in the design and features of photobioreactors to support controlled growth of photosynthetic microbes. The volume and capacity of photobioreactors are ranged from lab scale to industrial scale and the growth parameters are controlled tightly. Photosynthetic microbes can form a symbiotic association with non-photosynthetic microbes to enhance growth, nutrient recycling, and biomass productivity. This chapter focus on two important aspects of microalgae or cyanobacteria cultivation. The practical aspects of growth and photosynthetic physiology are discussed in detail along with description of latest advanced techniques and instruments for photosynthesis research.

4.2 Techniques to Assess Photosynthesis

4.2.1 Simultaneous Measurements of PSI and PSII Efficiency

Hand-held fluorometer, Infrared Gas Analyzer (IRGA), and Photosynthetic Efficiency Analyzer (Handy PEA) are a few utilitarian instruments that are effective in studying photosystem performance and other related parameters (Reddy et al. 2019; Sreeharsha et al. 2015). DUAL-Pulse Amplitude Modulator (DUAL-PAM 100, Walz, Germany) is another productive instrument for the simultaneous measurement of PSII-specific Chl *a* fluorescence signals (P680) and PSI-specific light absorption signals (P700). The utilization of a specialized module in the instrument enables the measurement of NADPH fluorescence in suspension cultures by the application of pulse-modulated UV-A excitation light. This feature offers significant benefits in NADPH measurements. The DUAL-PAM 100 was developed as a modular measurement system that allows for the measurement of a single pulse-modulated beam signal, typically Chl fluorescence excited by red or blue measuring light, as well as a dual beam difference signal, typically P700 absorbance detected through the difference in transmittance of 870 and 830 nm light. In addition, it is possible to measure two single beam signals as well. The emitter-detector units utilized for the measurement of chlorophyll fluorescence and P700 can be easily substituted with alternative emitter-detector units that have been specifically designed for the evaluation of various forms of optical modifications. These alternative units have the potential to offer supplementary insights into the process of photosynthesis. The NADPH detector module effectively addresses many limitations associated with NADPH fluorescence. NADPH exhibits fluorescence within the blue-green portion of the electromagnetic spectrum, with its highest excitation occurring in the UV-A region at approximately 340 nm and emission taking place at roughly 465 nm. The peak-to-trough amplitude of NADPH fluorescence is approximately 2000 times less than that of Chl fluorescence, and it is hindered by background signal originating from non-NADPH excitation sources, such as cellulose and flavins. The Dual PAM-100 is distinguished by its ability to seamlessly integrate user-friendly operation with an extensive array of potential applications in both fundamental and practical investigations pertaining to photosynthesis. Primarily, the apparatus measures various parameters such as light curves, slow and rapid kinetics, and other critical characteristics of light sources. These measurements encompass fluorescence excitation light as well as near-infrared (NIR) light.

4.2.2 Experimental Setup and Measuring Process

Power on the Dual-PAM 100 and allow it to initialize. Ensure that the instrument is calibrated and configured according to the manufacturer's instructions. Place the microalgae sample in a dark environment for sufficient *dark adaptation*. It can be done usually by covering the culture flask with black cloth or aluminum foil for sufficient time. The process of dark adaptation, which normally takes approximately 15 min at ambient temperature, ensures the closure of all photosystem II (PSII) and photosystem I (PSI) reaction centers, resulting in the stabilization of the F0 state.

Initial Fluorescence (F_0) *Measurement*: Attach or place the microalgae sample in the fluorometer chamber. Measure the minimal fluorescence (F0) by illuminating the sample with a weak modulated measuring light in the presence of a far-red actinic light that oxidizes the primary quinone acceptor (Qa) of PSII. *Measure Maximum Fluorescence (Fm)* Attach the microalgae sample to the instrument's sample holder or cuvette. Select the "Dark" mode or similar option on the instrument's software. Initiate a measuring sequence to record the dark-adapted fluorescence (F0). *Measure Maximum Fluorescence Yield During Light Exposure (Fm′)* by switching to the "Light" mode or equivalent setting on the instrument and thereby initiating a saturating pulse sequence (Table 4.1). *Measure Steady-State Fluorescence Yield During Light Exposure (Ft)* by keeping the instrument in the "Light" mode upon allowing the culture to acclimate to the measuring light intensity for a few minutes thereby initiating a measuring sequence. Record the calculated parameter values in a data sheet or software provided by the instrument. Repeat the measurements for multiple samples or under different experimental conditions if needed. Analyze the data to understand the photosynthetic performance of microalgae, including efficiency, photochemical and non-photochemical quenching, electron transport rates, and more.

The fluorescence parameters, specifically those related to Photosystem II (Fo, Fm, Fv, and Fm′) and Photosystem I (Po, Pm, and Pm′), offer significant insights on the efficiency and energy status of photosystems. The initial fluorescence emitted from dark-acclimated cells upon subjecting to the fluorimeter's low measuring beam is referred as F0. Following the application of a saturating pulse (SP), photosystem II (PSII) reaction centers result in the generation of the maximum fluorescence value (Fm′). The emission of preliminary fluorescence is significantly increased and stabilizes within a few seconds due to the surplus allocation of light energy to Photosystem II (PSII) by actinic light, which surpasses the capabilities of the Electron Transport Chain (ETC) (Kiran and Mohan 2021). Chapter 2 elucidated the movement of electrons originating from the activated Photosystem II (P680) to Photosystem I (P700) through a sequence of reducing agents, namely protein-bound plastoquinone A (QA) and subsequently plastoquinone B (QB). Similar pattern follows upon flashing actinic light on to the dark-adapted suspension culture. The process involves the dissociation of reduced hydroquinone (QBH2) from Photosystem II (PSII) and subsequent transportation of its reducing equivalents to Photosystem I (PSI). The process involves the reduction of oxidized P680 by a tyrosine residue found in the D1 subunit. This reduction then leads to the oxidation of a Mn_4CaO_5 cluster, which serves as the catalytic center responsible for the splitting of water (Bonnanfant et al. 2019). The light-harvesting complexes (LHCs) are responsible for the absorption of photons, which subsequently initiate linear and cyclic electron transfer (LET/CET) processes. These processes are essential for the conversion of CO_2 into storage carbons, while simultaneously releasing O_2 as a terminal product.

Table 4.1 Description of chlorophyll a fluorescence parameter and their interpretation

Parameter	Description	Interpretation	Formula
ABS/RC	Absorption Flux per Reaction Center	Reflects energy distribution within the photosynthetic apparatus	(FJ – F0)/(FM – F0)
TR0/RC	Trapped Energy Flux per Reaction Center	Reflects energy distribution within the photosynthetic apparatus	(FM – FJ)/(FM – F0)
ET0/RC	Electron Transport Flux per Reaction Center	Reflects electron transport rate through the open reaction centers	1/(2.9 * (FM – F0)) * (FJ – F0 + 4 * (FM – FJ))
DI0/RC	Dissipated Energy Flux per Reaction Center	Reflects dissipation of absorbed light as heat	1 – ABS/RC – TR0/RC
PI	Performance Index	Performance of the reaction centers when illuminated with light	(FJ – F0)/(FP – F0) * (FM – F0)/(FJ – F0) * (ET0/RC/TR0/RC)
φP0	–	Quantum yield of primary photochemistry	(FJ – FI)/(FJ – F0)
φE0	–	Quantum yield of electron transport	(FP – FJ)/(FP – F0)
φD0	–	Quantum yield of non-radiative energy dissipation	1 – φP0 – φE0
Fv/Fm	Photochemical yield	The efficacy of the photosystem II	(Fm – F0)/Fm
Y(II)			(Fm′ – Ft)/Fm′
ETR	Electron transport rate	Efficacy of photosystem to transport electrons	Y(II) * PAR * 0.84 * 0.5
NPQ	Non-photochemical quenching	Heat dissipation which reflects the wastage of absorbed light energy	(Fm – Fm′)/Fm′
qP	Photochemical quenching	Efficacy of the photosystem	(Fm′ – Ft)/(Fm′ – F0’)
qN	Non-photochemical yield	Effect of photoinhibition	(Fm – Fm′)/Fm
qT	Yield at specific time point		1 – Fv/Fm
Fv/Fo	Photochemical yield	Initial fluorescence	(Fm – Fo)/Fm

Non-photochemical Quenching (NPQ): NPQ is calculated based on the changes in the fluorescence yield during light exposure. The NPQ value at a specific time (t) can be calculated as follows:

$$\mathrm{NPQ}(t) = (\mathrm{Fm} - \mathrm{Fm}') / \mathrm{Fm}'$$

Where Fm: Maximum fluorescence in the dark-adapted state; Fm′: Maximum fluorescence at time t during light exposure.

4.2.3 OJIP Fluorescence Transients

A pulse amplitude magnitude fluorimeter (such as AquaPen-C 100) or (DUAL PAM 100) or Infrared Gas Analyzer (such as LiCor 6800) or any other similar equipment can be employed to measure the *Chl a* fluorescence transient. OJIP test is a series of transient chlorophyll fluorescence measurements and is a powerful analysis for studying the photosynthetic activity and energy transfer processes in suspension cultures. Otherwise known as chlorophyll fast kinetics, these parameters determine the structure and function of photosystems and assess quantum yield and electron transport in PSII under different experimental conditions (Kodru et al. 2015). The application of this technique includes intrinsic kinetic modeling, quantitative analysis of light and dark kinetics, measurement of photosynthetic rates, monitoring of photosynthetic capacity in photobioreactors, assessment of treatment effects on chlorophyll fluorescence, characterization of growth conditions and nutrient stress responses, evaluation of temperature optima for growth, identification of specific traits during nitrogen deprivation or carbon limitation. These studies contribute to the understanding of microalgal physiology and provide valuable insights for optimizing cultivation strategies.

To derive OJIP curves, the pulse-induced fluorescence is measured by subjecting dark-adapted cells to an illuminating pulse with a saturating pulse amplitude of 3000 mol m^{-2} s^{-1}. The detection of fluorescence intensity occurs over a time range spanning from 11 μs (F0) to 1 s. Over this range, specific time points are designated as L, K, J, I, and P/Fm bands, corresponding to intensity measurements at 150, 300, 2 ms, 30 ms, and 500 ms to 1 s, respectively. These bands are utilized for the computation of OJIP test kinetics, which provides vital insights into the functioning of Photosystem II (PSII) (Ranadheer et al. 2019).

O: F0 to FJ transition (from open to closed PSII centers).

J: FJ to FI transition (from intermediate to fully closed PSII centers): Apply a saturating light pulse to close all PSII reaction centers (transition from open to closed state) and measure the fluorescence value (FJ) during the J-step.

I: FI to FP transition (from fully closed to partially reduced PSII centers).

P: FP to FM transition (maximum fluorescence): Apply a saturating light pulse to achieve the maximum fluorescence level and measure the maximum fluorescence value (FP) during the P-step.

The O–J–I–P and S–M–T transients are utilized to evaluate fluorescence induction when exposed to saturating light. These transients demonstrate a rapid multiphasic increase in the microsecond (μs) to second (s) time range. "K band," "L band," "I band," and "P band" refer to specific wavelength ranges used in measuring and analyzing the chlorophyll fluorescence response of microalgae (Fig. 4.1).

K Band: The K band corresponds to the initial segment of the OJIP curve. It represents the period when the fluorescence signal rises rapidly from its minimal level (O) to a peak (J) in response to a saturating light pulse. The observed increase in value is suggestive of the decline in the major electron acceptor, QA, within the reaction center of photosystem II (PSII). The K band provides information about the efficiency of energy capture and electron transport in PSII.

L Band: The L band follows the K band and represents the segment of the OJIP curve between points J and I. In this phase, the fluorescence signal continues to rise, but at a slower rate compared to the K band. The L band reflects the reduction of the plastoquinone (PQ) pool and the electron transport beyond QA, including the movement of electrons through the cytochrome b6f complex and the PQ pool. This phase provides insights into the efficiency of electron transport downstream of QA.

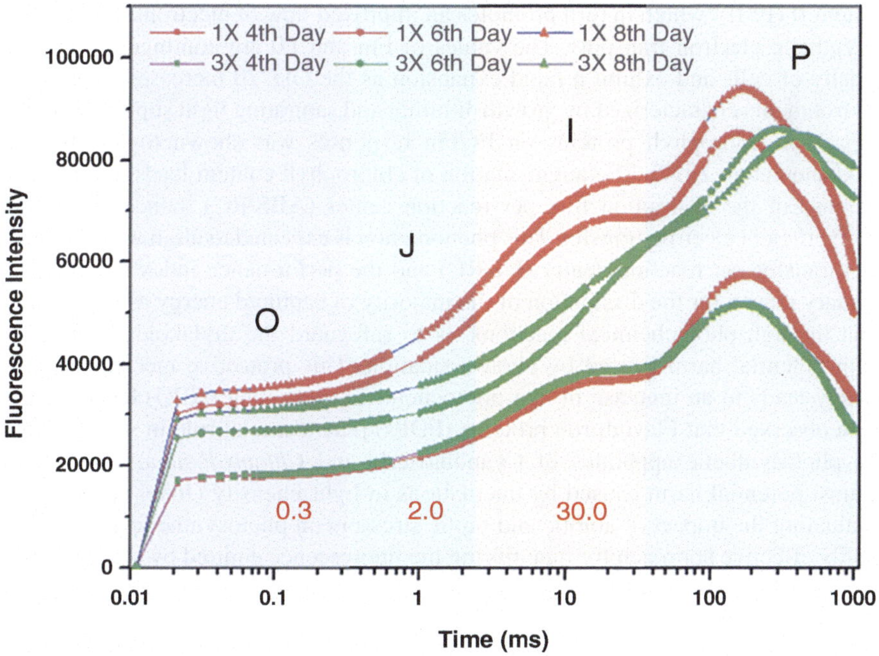

Fig. 4.1 Typical OJIP curves measured in microalgae cultures representing O, J, I, and P phases. (Reprinted with permission from Ranadheer et al. 2019)

I Band: The I band corresponds to the point of inflection between the J and P phases of the OJIP curve. This point is often associated with the reduction of the plastoquinone (PQ) pool and the electron transport downstream of QA. The I band provides insights into the efficiency of electron transport beyond QA and before the system reaches its maximum fluorescence level.

P Band: The P band corresponds to the peak fluorescence level (FP) reached after the I phase. It indicates the maximum fluorescence level that the photosynthetic system can achieve under the given conditions. The P band is important for assessing the overall photosynthetic performance and the capacity of the photosynthetic machinery.

Photosystem II (PSII) employs the cytochrome (Cyt) b6f complex to perform the oxidation of water and the reduction of plastoquinone (PQ). On the other hand, Photosystem I (PSI) carry out the oxidation of plastocyanin (PC) and the reduction of ferredoxin. The increase in OJ fluorescence during the photochemical phase is associated with the reduction of the primary electron acceptor, QA, to its reduced form, Q-A. Conversely, the thermal phases known as JI and IP are responsible for the removal of the PQ pool and the electron acceptor side of PSI, respectively (Tikhonov 2014). The occurrence of SM rise transients can be attributed to the inactivation of LHC II kinase, which takes place when LHC IIs undergo dephosphorylation and then migrate from the PSI to the PSII region as a result of exposure to high-intensity light. A higher Fv/Fm ratio signifies the presence of an efficient oxygen-evolving complex (OEC) as well as electron donors and acceptors at photosystem II (PSII), which in turn promotes an improved flow of electrons during photosynthetic electron transport. The values of Fm and F0 are contingent upon the density of cells and exhibit a rapid expansion as the OD720 increases. In cellular environments characterized by growth-limiting and saturating light suppression, the impact of chlorophyll proteins on Fv/Fm responses was shown to be minimal (Tikkanen et al. 2017). The augmentation of chlorophyll content leads to a notable increase in the absorption flux per reaction center (ABS/RC), hence promoting more efficient electron transfer. This phenomenon is expected to align with the electron transfer per reaction center (ET/RC) and the performance index (PI_{ABS}). The primary reason for the dissipation of the majority of captured energy as heat, rather than through photochemical reactions, is to safeguard the thylakoid membranes from potential harm caused by photo-oxidation. This protective mechanism ultimately leads to an increase in non-photochemical quenching (NPQ) levels. It has been observed that Flavodiiron proteins (FDPs) play a crucial role in safeguarding the photosynthetic apparatus of Cyanobacteria and *Chlamydomonas reinhardtii* against potential harm caused by fluctuations in light intensity (Jokel et al. 2018). Evaluating the impact of abiotic and biotic stressors on photosynthetic activity is a highly effective approach for quantifying the fluorescence emitted by specific components of Photosystem II (PSII) due to the excitation of chlorophyll molecules (Kodru et al. 2015).

The light peak absorption levels of Photosystem II (PSII) and Photosystem I (PSI) exhibit distinct differences; however, the activation of Light-Harvesting Complex II (LHCII) trimers confers advantages to both photosystems. The YPSII metric is a quantitative assessment of the quantum yield associated with electron transport in photosystem II (PSII). It takes into account the excitation of electrons to initiate PSII activity, as well as the subsequent reopening of PSII after a photochemical event. *Tetradesmus* Sp. exhibited elevated F0 and Fm values, resulting in a higher Fv/Fm ratio, which suggests efficient electron donation and acceptance capabilities. As the actinic light (AL) intensity increased, the saturating pulse-induced fluorescence under AL (Fm′) exhibited a significant decrease and was then normalized to the maximal fluorescence of cells acclimated to darkness (Fm). This normalization resulted in the calculation of the Fm′/Fm metric (Tikkanen et al. 2017). ETRII is a quantitative measure of electron flow within the photosynthetic cycle, which exhibits a strong correlation with the process of oxygen release and carbon dioxide absorption. Higher Fv/Fm ratios at Photosystem II (PSII) are indicative of efficient water splitting and carbon fixation, facilitating enhanced electron transport during the process of photosynthesis. The acceleration or deceleration of PSII and ETR(II) is contingent upon the physiological condition and can exhibit hysteresis in response to short- and long-term irradiance variations. The dissipation of excess energy in microalgae occurs through passive mechanisms, namely heat dissipation and fluorescence emitted by light-harvesting complexes. These complexes play a crucial role in the production of Y(NO) and Y(NPQ). The Fv/Fm ratio experienced a significant decrease when subjected to intense light throughout the developmental phase, suggesting the occurrence of non-photochemical quenching and subsequent inactivation of the photosystem II (PSII) complex, also referred to as photoinhibition. The variable Y(NO) has an inverse relationship with YPSII. The unregulated energy dissipation, denoted as Y(NO), remained consistent and ineffective in promoting thermal dissipation during the progression of development. The alteration in Fm and Fm′ quantifies the phenomenon of non-photochemical quenching (NPQ). The PSII percentage was reduced to 0.16 following exposure to a light intensity of 538 μmol m^{-2} s^{-1}, indicating that the yield of photosystem II (PSII) was sufficiently diminished. The induction of Y(NPQ) and Y(NO) resulted in a slight decrease in the expression of PSII. The quantification of energy loss is contingent upon several factors, including the specific species of algae, its growth rate, nutrient availability, light exposure, and ambient temperature (Wimalasekera 2019).

4.2.4 P700 and Cyt b6f studies

The P700 signal, which is determined by measuring the variance in absorbance at wavelengths of 875 and 820 nm, offers supplementary insights into the regulation of photosynthesis (Zivcak et al. 2015). By oxidizing the ETC with PSI-favoring far-red light and then adding an SP on top, PM is achieved. The portion of P700 that was oxidized by the actinic light is designated as Y(ND), which stands for the donor side limitation of PSI. Y(ND) is inversely correlated with Y(I). The pH is necessary for Cyt b6f to control electron transport during photosynthesis, and high PSI stimulation lowers Y(ND). The PSI acceptor side limit, Y(NA), is controlled by the intersystem ETC's redox state. Oxygenic photosynthesis, which takes place in the thylakoid membrane, captures sunlight and divides water to create ATP and decrease equivalents that are crucial for CO_2 uptake and cell metabolism. The effective quantum yield (Fv/Fm) of PSII and PSI progressively drops with increasing light intensity, principally as a result of a fall in GP thus increasing the NPQ. The Y(ND) and Y(NA) parameters are especially useful in indicating the donor and acceptor side reduction status of the PSI. While low light intensity results in little cyclic electron flow activation (CEF) which can be understood from the values of ETRI and ETRII. CEF was involved in the photosynthetic CO_2 assimilation, photoprotection of the OEC, and defense of PSI against oxidative damage (Huang et al. 2019). The promotion of photosynthetic assimilation in chloroplasts during treatment is facilitated by high electron transport and the establishment of a trans-thylakoid proton gradient, resulting in the production of ATP and NADPH (Jokel et al. 2015). The parameter Fv/Fm is influenced by the processes of PSII/PSI photoinhibition and the dark oxido-reduction of the plastoquinone pool. Photoinhibition usually hinders the efficiency of photosynthesis by impeding the formation and degradation (turnover) cycle of the light-harvesting complex, which involves majorly the D1/D2 proteins, located at Photosystem II (PSII) (Chiranjeevi and Mohan 2016). When the cellular metabolic capacity is surpassed by light energy, the mechanism reliant on the PGR5 protein induces an elevation in pH levels (Kiran and Mohan 2021). The regulation of electron transfer through the cytochrome b6f complex (Cyt b6f) and the light-harvesting efficiency of the thylakoid antenna system is governed by the trans-thylakoid proton gradient (ΔpH). The activity of the pH-sensitive PSBS protein is contingent upon the pH level, whereby it facilitates the dissipation of surplus excitation energy in the form of heat. The observed elevation in pH, along with a decrease in YII within the PSII complex, suggests a potential issue with the process of electron transport (Malapascua et al. 2014).

4.3 Harvesting Techniques

The downstream processing of microalgae is an essential technique that encompasses the acquisition, modification, extraction, and purification of particular components with high value. The cost of harvesting in microalgal biorefineries might vary between 20% and 30% depending on the specific variety of microalgae and the technology utilized. An efficient resource optimization and harvesting mechanism, encompassing centrifugation, filtration, flocculation, and floatation or sedimentation techniques will result in a reduction in the overall cost of the downstream processing stage in a microalgal biorefinery (Yin et al. 2020). In comparison to alternative techniques, flocculation can be regarded as a highly cost-effective and economically viable procedure. Flocculation is a technique employed to improve the effectiveness of microalgal sedimentation through the utilization of flocculating chemicals. Cell disruption is a crucial downstream process that contributes to the increased cost of the entire microalgal biorefinery. This cost escalation is mostly due to the utilization of cell disruption devices, including homogenizers, bead pounding, French presses, chemicals, and high-pressure heating (Jacob-Lopes et al. 2015). The designated biomass conversion procedures should be conducted in a manner that does not have any negative effects on the other components. Additionally, the advancement of downstream processing technologies will provide substantial advantages in terms of processing efficiency and cost considerations. The biomass has the potential to be collected and utilized as a nutrient source or incorporated into various organic commodities. The post-lipid extraction processing of microalgal biomass plays a crucial role in the generation of value added compounds derived from microalgae. The procedure encompasses a series of sequential stages, including the collection of raw materials, the removal of excess moisture, the desiccation process, and the separation into distinct components. The microalgae that have been collected are segregated into distinct fractions and subjected to various methodologies, including solvent extraction, supercritical fluid extraction, and enzymatic hydrolysis, for further processing. Following the process of lipid extraction, the residual biomass can be repurposed for several alternative applications, including but not limited to animal feed, biofuel production, or fertilizer manufacture. The utilization of processed microalgae biomass also presents opportunities for the extraction of valuable nutrients or bioactive substances. In order to achieve optimal product quality, it is imperative to meticulously plan and oversee the post-lipid extraction processing of microalgae biomass.

According to studies, sodium carbonate supplementation significantly enhanced both lipid and biomass content whereas nitrogen depletion raised lipid content and delayed development (Sivaramakrishnan and Incharoensakdi 2017). The high CO_2 fixation efficiency of an artificial light source was found to enhance biomass content more than direct sunshine, according to a study on the effects of light intensity on microalgal growth (Aravantinou and Manariotis 2016). However, the cost of artificial light can be offset by the cultivation of huge biomass and upgraded high-value goods. The primary component that has a significant impact on microalgal growth is light intensity. Yet it is evident that nutrition and light are crucial for algal growth.

Different techniques, including solvent extraction, microwave-assisted extractions, electroporation, and ultrasonic extraction, can be used to extract microalgal lipids. The principal drawbacks of lipid extraction, however, include their high energy requirements, extensive solvent prerequisites based on the type of lipids found in the microalgae, relatively high temperatures, and high flammability hazards (Hernández et al. 2014).

Recently, supercritical extraction methods were taken into consideration to substitute hazardous solvents; the main benefit of this method is that it exhibits excellent selectivity towards triglyceride, an essential molecule needed for the manufacture of biodiesel. For the supercritical extraction of lipids, CO_2 is frequently used since it is safe, affordable, environmentally benign, and recyclable. One primary drawback associated with employing a supercritical technique is its substantial energy consumption, which exerts a notable influence on the overall process. Novel methodologies, including isotonic extraction, osmotic pressure extraction, and enzyme-based extraction, have been developed for the purpose of lipid extraction, eliminating the need for solvent usage. Due to the lack of a solvent, solvent-free processes are thought of as being easier and more straightforward than solvent methods. The biomass that was used to extract the oil (also known as "waste biomass") can also be used to make ethanol or animal feed (Ho et al. 2014).

4.4 Culturing Techniques

4.4.1 Two-Stage Cultivation of Microalgae

The two-stage cultivation approach involves a two-stage cultivation method, wherein the initial stage is focused on achieving optimal growth conditions to maximize biomass output. Subsequently, the second stage is dedicated to the accumulation of lipids or other value-added products, employing different stress conditions. Two-stage cultivation of microalgae is a strategy used to optimize the growth and productivity of microalgae for various applications, such as biofuel production, bioremediation, and high-value compound extraction. This approach involves two distinct phases of growth under different conditions to enhance biomass and product yields. The two stages typically include the growth stage and the production stage.

Growth Stage: In the first stage, microalgae are cultivated under conditions that promote rapid biomass accumulation. This usually involves providing optimal nutrients (nitrogen, phosphorus, and trace elements), light, and temperature. The goal of this stage is to maximize cell division and growth rate, resulting in a high cell density culture. However, growth conditions might not be optimal for the production of specific target compounds.

Production Stage: After reaching a desired biomass concentration in the growth stage, the microalgae culture is transferred to the production stage. In this stage, the cultivation conditions are manipulated to induce the synthesis and accumulation of specific target compounds, such as lipids for biodiesel, pigments, or other bioactive molecules. This might involve stressing the microalgae by altering nutrient availability, light intensity, temperature, or other factors.

4.4.1.1 The Benefits of a Two-Stage Cultivation Approach

Optimized Biomass and Product Yields: By separating the growth and production phases, microalgae can be cultivated under conditions specifically tailored for each phase, resulting in higher overall productivity of both biomass and target compounds.

Controlled Production Conditions: The production stage allows for tighter control over the environmental conditions that influence the synthesis of specific compounds. This can lead to higher yields of the desired products.

Reduced Competition for Resources: In a two-stage system, resources like light and nutrients can be allocated more efficiently, as they are directed towards either rapid growth or product accumulation in separate phases.

Flexibility: This approach provides flexibility to optimize different parameters for each stage, enabling researchers to adapt to various microalgae species and target compounds.

Economic Viability: Separating growth and production stages can lead to cost savings, as it reduces the need for continuously maintaining optimal conditions for both growth and product accumulation.

4.4.2 Cocultivation: Algal—Bacterial—Consortia

Algae are known for their ability to recover nutrients by depleting the available inorganic compounds. In natural ecosystems or culture conditions, the efficiency of this process is often stimulated by the co-occurrence of bacteria in the environment. Particularly, in wastewater treatment systems, the interactions between these diverse microbes bring synergistic benefits of cost-effective aeration, pollutant removal, and greenhouse gas emission sequestration. Bacteria exploit the oxygen generated as a result of algal photosynthesis, whereas algae absorb the carbon dioxide emitted by bacteria. Additionally, algae exhibit pathogen-removal capabilities, including viruses thus making the relationship interdependent on each other (Kouzuma and Watanabe 2015; Saravanan et al. 2021). The availability of rich bacterial population in aquatic ecosystems promotes a natural partnership with algae, functioning as a water purification mechanism thus forming a foundational and complex aspect of ecosystems. Nonetheless, such symbiosis poses a challenge in the case of biorefineries as separation of algal and bacterial biomass which are intact together in the liquid cultures (Ramanan et al. 2016). Recent studies on wastewater treatment have reported formation of clusters composed of algae and bacteria facilitating the downstream biomass processing through sedimentation, without flocculating agents (Ryu et al. 2017; Shen et al. 2017). Other than liquid cultures, immobilized algal-bacterial cells are also deployed in immobilized format to treat municipal wastewater. In the treatment of saline wastewater, marine microalgal-bacterial consortia have been utilized in fixed bed photobioreactors (Babatsouli et al. 2015), while algal-bacterial biofilm bioreactors have been employed for domestic wastewater sludge treatment (Posadas et al. 2013). The treatment of pharmaceutical wastewater involves the removal of compounds by algal-bacterial consortia through processes such as bioaccumulation or biodegradation. In instances of metal-containing wastewater, heavy metals are eliminated through biosorption, bioconversion, and bioaccumulation (Wang et al. 2016). Algal-bacterial symbiosis and their metabolic interactions are majorly beneficial in terms of nutrient exchange, gene transfer, and signal transduction. The prominent co-cultivation demonstrations between microalgae and other microbes are listed in Table 4.2.

4.5 Photobioreactors

Photobioreactors are artificially constructed cultivation system which facilitates the growth of light-dependent photosynthetic microbes under optimal conditions of light, temperature, pH, and nutrients. They are made of acrylic or glass materials thus enabling the easy passage of light and aids in photosynthesis. In order to mimic the natural light in indoor conditions, photobioreactors are fitted with light emitting diodes of desired light intensity. Under favorable light intensity and controlled growth conditions, PBRs can achieve algal yields up to 100 $g/m^2/h$ which is several folds higher than open cultivation. PBRs can be configured into different dimensions and shapes including tubular, flat plate/flat panel, or vertical column type structures in which the microbial cultures can be pumped and recirculated continuously (Carvalho et al. 2013; Ullah et al. 2015). Nevertheless, PBRs are much costlier than open pond systems, but their application in industrial biorefineries are justified in targeted applications because of advantages. The systems in question are completely enclosed, thereby effectively preventing contamination from external sources such as foreign algal species, fungi, or amoeba. Unlike open and covered ponds, these closed systems do not experience water evaporation losses. Additionally, they possess improved mechanisms for nutrient dispersion and heat dissipation, resulting in a more uniform and controlled growth of algae biomass. Furthermore, growth in PBRs provides an opportunity to control and monitor nutrient, CO_2 levels. Usage of LEDs facilitates photosynthesis even during night conditions thus enhancing the growth and biomass productivity of algal cultures. Selection of type of photobioreactor depends on growth condition and species type as they demand specific growth condition. Various scales of cultivation reactors for microalgae growth are represented in Fig. 4.2.

4.6 Airlift Reactors (ALRs)

ALRs possess a high degree of flexibility and demonstrate significant suitability for integration into biorefineries and integrated power generation systems. This is primarily down to the wide range of process parameters available, which may be effectively utilized to enhance their operational performance. Microorganisms are deemed superior to tubular bubble reactors and open ponds for their growth. This is primarily due to the homogeneous shear distribution throughout the ALR (algae cultivation systems with light) which results in a relatively constant environment. Consequently, this leads to the minimization of abrupt fluctuations in the mechanical forces exerted on suspended particles (Merchuk and Camacho 2010). Furthermore, ALRs offer the advantage of achieving rapid photoperiods and efficient mixing without requiring excessive energy consumption. The interest in ALRs for the cultivation of microalgae is particularly heightened due to their ability to achieve high yields while minimizing energy consumption, especially in cases of medium to low mixing.

Table 4.2 Summary of different cocultivation strategies adopted for microalgae and cyanobacterial cultures

Microalgae species	Co-cultivated species	Mode of cultivation	Outcome	References
Chlorella minutissima	*Aspergillus awamori* (fungus)	Mixotrophic	Enhanced biomass and lipid production	Dash and Banerjee (2017)
Nannochloropsis oculata	*Aspergillsus fumigatus* (fungus)	Photobioreactor	Enhancement of total biomass, lipids production, and optimization of fatty acids composition	Wrede et al. (2014)
Chlorella vulgaris	*Ganoderma lucidum* (fungus)	Mixotrophic	Upgraded biomass, and biogases purification has been done	Cao et al. (2017)
Chlorella vulgaris	*Streptomyces rosealbus* (bacteria)	Photobioreactor	Enhanced biomass and lipid productivity	Lakshmikandan et al. (2021)
Chlorella pyrenoidosa	*Rhodobacter capsulatus* (bacteria)	Mixotrophic	Increased in biomass, acetate assimilation rate, and lipid productivity	Shen et al. (2022)
Chlorella saccharophilla	*Cellvibrio pealriver* (bacteria)	Photobioreactor	Enhanced lipid production	Xie et al. (2021)
Botryococcus braunii	*Rhizobium sp.* (bacteria)	Mixotrophic	Hydrocarbon for biofuel	Rivas et al. (2010)
Scenedesmus sp.	*Azotobacter vinelandii* (bacteria)	Mixotrophic	CO_2 fixation without fertilizer added	Villa et al. (2014)
Chlorella protothecoides	*Chlorella protothecoides* (microalgae)	Mixotrophic	Lipids for biodiesel, lutein for food	Santos et al. (2014)
Chlorella sp.	*Ettlia* sp. (microalgae)	Mixotrophic	C16–C18-rich fatty acid production	Rashid et al. (2019)
Chlorella sp.	*Monoraphidium* sp. (microalgae)	Photoautotrophic	Enhanced lipid productivity	Zhao et al. (2014)
Haematococcus pluvialis	*Phaffia rhodzyma* (yeast)	Mixotrophic	Astaxanthin production	Dong et al. (2006)
Spirulina platensis	*Rhodotorula glutinis* (yeast)	Mixotrophic	Lipids for biodiesel	
Chlorella vulgaris	*Rhodotorula glutinis* (yeast)	Mixotrophic	Enhanced lipid content	Zhang et al. (2014)

(continued)

Table 4.1 (continued)

Microalgae species	Co-cultivated species	Mode of cultivation	Outcome	References
Scenedesmus obliquus	Rhodotorula glutinis (yeast)	Mixotrophic	Increased biomass and lipid yield	Yen et al. (2015)
Chlorella pyrenoidosa	*Rhodotorula toruloides* (yeast)	Mixotrophic	Improved both lipid production and wastewater treatment efficiency	Ling et al. (2014)
Tribonema sp.	*Saccharomyces cerevisiae* (yeast)	Photobioreactor	Enriched lipid and carbohydrate content	Wang et al. (2014)

4.7 Tubular Bioreactors

Horizontal tubular bioreactors consist of transparent tubing that is oriented horizontally, and they are equipped with integrated mechanisms for gas transport. Helical tube bioreactors are designed with flexible transparent tubing arranged in a helical manner, allowing for the accommodation of longer tubing lengths within a limited space. Consequently, these bioreactors exhibit notable spatial efficiency (Chen et al. 2011). The scaling up of vertical tubular reactors necessitates the consideration of increased working volume. This entails the use of pipes or tubes with larger diameters, which in turn compromises the surface to volume ratio and leads to a drop in photosynthetic efficiency. One significant limitation associated with these bioreactors is their tendency to reflect sunlight as a result of their substantial angular orientation (Carvalho et al. 2006). Nevertheless, the issue was resolved by implementing horizontal tubular bioreactors. However, it should be noted that the system's high light penetration intensity sometimes leads to the negative of increased heating. Elevated temperatures impede the proliferation of algae, necessitating the implementation of supplementary cooling mechanisms. Various cooling technologies are employed, such as air cooling and the application of water onto the tubes. However, it should be noted that these methods contribute to an increase in water usage. Helical tubular bioreactors are a viable alternative to horizontal tube bioreactors. The system comprises a framework containing a coiled polyethylene tube, which is flexible in nature. This tube is connected to a specific gas exchange unit and a heat exchange unit. Additionally, a centrifugal pump is affixed to the system to facilitate effective mixing of the culture broth.

4.8 Flat Plate Bioreactors

The primary objective of these bioreactors is to optimize solar energy utilization for enhanced efficiency. The system primarily comprises many flat modules that are interconnected either in series or parallel, with the inclusion of an optional degassing unit. Therefore, these flat plates significantly enhance the ratios of surface area to volume. One of the primary drawbacks associated with flat plate systems is the disruption of flow control. Due to the notably high photosynthetic efficiency, growth rates are substantially elevated, resulting in a substantial increase in the oxygen concentration within the culture. Consequently, the implementation of a degasser becomes necessary. This degasser can be either a closed or an open system, as several innovations have been introduced in this regard (Sreeharsha and Venkata Mohan 2021).

4.9 Open/Raceway Ponds

"High-rate algal pond (HRAP)" are developed based upon oxidation pond technology in the mid of twentieth century to treat wastewater with minimal inputs. HRAP are shallow, paddle wheel, raceway style, and mixed lagoon-based system. Compared to conventional wastewater treatment technologies HRAPs are sustainable, cost-effective, and operation costs typically 70% less than the activated sludge system. HRAPs have been used with different wastewaters viz., domestic, industrial, municipal, and livestock for nutrient removal (Mohan et al. 2011, 2015; Nie et al. 2020; Sutherland et al. 2017). Within HRAPs, heterotrophic bacteria play a crucial role in the decomposition of organic materials, resulting in the production of carbon dioxide. This carbon dioxide is essential for algal photosynthesis, which subsequently provides the oxygen necessary for aerobic respiration (Robles et al. 2020). Microalgae cultivation in wastewater reduces the production cost, mitigates carbon dioxide, and enhances nutrient removal through assimilation of organic compounds into the biomass with wide applications as cattle and aquaculture feed, soil conditioner, and production of biofuels (Hemalatha et al. 2019; Tang et al. 2020). Microalgae cultivation in freshwater HRAPs requires 40 to 100 kg of Nitrogen, 3 to 12 kg of Phosphorous, 11–13 ML/ha/year of water, and other micronutrients to produce 1 ton of algal biomass. In contrast, wastewater utilization in algal ponds requires no extra nutrient addition. Algal cultivation in wastewater provides a niche for the production of low-cost biofuel (Park et al. 2013; Rohit and Venkata Mohan 2018). Despite HRAPs is a well-known technology, wastewater treatment and biomass productivity are hindered by physicochemical and biological factors within the pond. Photosynthesis plays a crucial role in wastewater treatment and factors influencing the fluctuation in HRAP are light, temperature, pH, nutrient load, strain selection, carbon availability, hydraulic retention time, and algal recycling. Light in HRAP varies with diurnal irradiance and appears to be the major controlling factors for microalgae photosynthesis. High biomass concentration and suspended particulates in the wastewater increases light attenuation in HRAPs and leads to decrease in photosynthesis and nutrient recovery. Microalgae traps atmospheric sunlight using light-harvesting antennae complexes and transferred to the photosystem reaction centers where photochemistry takes place. Higher light intensities cause photoinhibition and reactive oxygen species (ROS) generation in plasma membrane which further inhibits photosynthesis, proteins, and cell membrane lipids. Under light diminishing conditions, chlorophyll efficiency decreases within the cell due to self-shading.

Carbon plays a crucial role in microalgae photosynthesis and constitutes 50% of its organic biomass. Unlike plants microalgae absorbs photons and utilize light energy for CO_2 assimilation and formation of carbohydrate moieties (Uggetti et al. 2018). In HRAPs, cultivation of heterotrophic microalgae requires 25–50% of inorganic carbon. Low carbon/nitrogen ratio in wastewater limits removal of nutrients and microalgae production in HRAPs compared to algae-bacterial biomass ratio (Caia et al. 2018; De Godos et al. 2016). Elevated pH decreases CO_2 and affects dissolved inorganic carbon in HRAPs. pH in HRAPs increases with photosynthetic reduction of carbon during day and decreases respiration at night (Craggs et al. 2012). Alkaline pH (>9) affects microalgae photosynthesis by reducing the available CO_2, ceases utilization of HCO_3^- and CO_3^{2-}, and interferes with RuBisCO catalyzing carboxylation of RuBP and photorespiration. An elevated pH helps in the formation of free ammonia and under high light intensities it inhibits photosynthesis, membrane transport, and nutrient uptake and accumulation of toxic elements (Clark and Flynn 2000). High pH has a negative effect on wastewater treatment by inhibiting the bacterial growth and oxidation of organic material to CO_2.

Microalgae grow over a wide range of temperatures and the optimum temperature required for photosynthesis is in between 28 and 35 °C (Ras et al. 2013; Subhash et al. 2014). Below the optimal or increase in temperature will enhance the photosynthetic pigments, cytosolic enzymes, biomass yield, and alter the lipid profile. This not only increases cell energy but the proportion of saturated to unsaturated fatty acids increases that are beneficial for production of biofuels (Wu et al. 2013). The essential macronutrients required for microalgae growth are nitrogen and phosphorus. Change in N:P ratio affects nutrient removal efficiency and has negative effects on microalgae growth by reducing photosynthesis (P_{max} and NPQ) and inhibits cell division (Devi et al. 2012). In wastewater HRAPs, under controlled conditions phosphorous rarely limits algal growth whereas nitrogen limits growth of algae. Under nitrogen limitations, protein synthesis declines while lipid and carbohydrates increase whereas under phosphorous limitation, membrane phospholipids are replaced by non-phosphorous lipids useful for biofuel production (Buchanan et al. 2018). In HRAPs, Cladocera and Rotifers taxa grazers enter the pond from the surrounding environment. At higher concentrations, these taxa severely reduce the algal biomass, dominance, modify size and shape of colonies. They induce protective spines that affect microalgal settleability (Park et al. 2011) (Fig. 4.2).

Challenges faced by HRAPs can be mitigated through modification of pond design, strain selection, hydraulic retention time, CO_2 addition, and operation cost.

Fig. 4.2 Culturing techniques for microalgae representative figure showing different methods of operating liquid cultures of photosynthetic microbes. a) multicultivator b, c) Flask cultures d) Photobioreactor e) Flat plate bioreactor f) Extracted microalgal biomass

These factors influence microalgae cultivation with high productivity and significant nutrient removal. Microalgae species capable of removing nutrient load and with desirable attributes like elevated photosynthetic rate, CO_2 sequestration, and centralizing carbon source could enhance wastewater treatment in HRAPs. Green algae from genera *Pediastrum*, *Desmodesmus*, *Micractinium*, and *Mucidosphaerium* present in wastewater HRAPs are tolerant to wide environmental variables (Mehrabadi et al. 2017). Algal species with small chlorophyll antenna size and genetically insertion of accessory pigments (phycobilins) helps in improving light absorption and increases biomass yield in HRAP (Simionato et al. 2013). Increase in enzymatic kinetics and CO_2 specificity can be achieved by introducing genetically engineering RUBisCO into algal cells.

Temperature fluctuations and light availability in shallow ponds affect the microalgae productivity compared to deep ponds. In euphotic zone, vertical mixing ensures light equal exposure to all the algae cells thereby promoting enhanced growth, photosynthetic activity and minimizes thermal stratification (Vadiveloo and Moheimani 2018). It reduces the boundary layer, nutrient assimilation, gas exchange and prevents formation of sediments at bottom of the pond. In HRAPs, rate of mixing (circuit time) is directly proportional to the colony size and harvest method. Large colonies and flocs are preferred over single-celled algae due to ease of harvesting (Guo et al. 2020).

Hydraulic retention time (HRT) plays a crucial role in modifying biomass concentration, algae/bacterial ratio, and wastewater treatment in HRAP with respect to energy (Park et al. 2011; Robles et al. 2020). Retention time can be configured by varying nutrient load, pond depth, and change in biomass concentration depending on season and altitude. Pond depth ensures effective delivery of light to each algal cell and governing biomass concentration. HRAPs have optimum depth which ranges from 15 to 100 cm (Grobbelaar 2013). Shallow ponds have high biomass concentration whereas high nutrient removal and aerial productivity was achieved by deeper ponds (Nie et al. 2020). Most of the land on earth surface is predominantly occupied by agricultural crops; operating a greater number of shallow ponds are not economically feasible. So, designing wastewater HRAPs with increased depths to retrofit in existing facultative ponds will increase microalgae photosynthesis and enhance nutrient removal. CO_2 addition brings biological changes in algae which enhances photosynthetic process and light absorption capacity thereby reducing the package effect. Addition of CO_2 to HRAP brings additional inorganic carbon source for microalgae growth under carbon limitations (Yong et al. 2021). In wastewater HRAP, reduced pH resulted in an increase in algal biomass and nitrogen availability by 30% and 17%. CO_2 addition also enhances algae energy content and fatty acid content (Posadas et al. 2015).

4.10 Conclusion

Assessment of photosynthetic efficiency and optimizing the growth conditions are two crucial aspects in suspension cultures of photosynthetic microbes. In recent times, advanced fluorimeters are available to understand the responses of photosystems to various environmental conditions. This ultimately determines the growth and productivity of microbes. Infrared gas analyzer, Dual PAM, and AquaPen are few instruments which are highly useful in measuring fluorescence transients of photosystems. The present chapter also discusses different culturing techniques and strategies for microalgae cultivation, along with their benefits, challenges, and applications, with a focus on their role in biofuel production and wastewater treatment. Mixotrophic culturing of microalgae in photobioreactors needs to be explored at an industrial scale in order to get cost-effective and sustainable biomass for diverse applications.

References

Aravantinou AF, Manariotis ID (2016) Effect of operating conditions on Chlorococcum sp. growth and lipid production. J Environ Chem Eng 4:1217–1223

Babatsouli P et al (2015) Single stage treatment of saline wastewater with marine bacterial–microalgae consortia in a fixed-bed photobioreactor. J Hazard Mater 292:155–163

Bonnanfant M et al (2019) Photosynthetic electron transport transients in Chlorella vulgaris under fluctuating light. Algal Res 44:101713

Buchanan NA et al (2018) Performance of a high rate algal pond treating septic tank effluent from a community wastewater management scheme in rural South Australia. Algal Res 35:325–332

Caia M et al (2018) Optimizing CO_2 transfer in algal open ponds. Algal Res 35:530–538

Cao W et al (2017) Simultaneously upgrading biogas and purifying biogas slurry using cocultivation of Chlorella vulgaris and three different fungi under various mixed light wavelength and photoperiods. Bioresour Technol 241:701–709

Carvalho AP et al (2006) Microalgal reactors: a review of enclosed system designs and performances. Biotechnol Prog 22:1490–1506
Carvalho F et al (2013) Cheese whey wastewater: characterization and treatment. Sci Total Environ 445:385–396
Chen C-Y et al (2011) Cultivation, photobioreactor design and harvesting of microalgae for biodiesel production: a critical review. Bioresour Technol 102:71–81
Chiranjeevi P, Mohan SV (2016) Critical parametric influence on microalgae cultivation towards maximizing biomass growth with simultaneous lipid productivity. Renew Energy 98:64–71
Clark DR, Flynn KJ (2000) The relationship between the dissolved inorganic carbon concentration and growth rate in marine phytoplankton. Proc R Soc Lond Ser B Biol Sci 267(1447):953–959
Craggs RJ et al (2012) Wastewater treatment and algal biofuel production. In: Algae for biofuels and energy. Springer, Dordrecht, pp 153–163
Dash A, Banerjee R (2017) Enhanced biodiesel production through phyco-myco co-cultivation of Chlorella minutissima and Aspergillus awamori: an integrated approach. Bioresour Technol 238:502–509
De Godos I et al (2016) Evaluation of high rate algae ponds for treatment of anaerobically digested wastewater: effect of CO_2 addition and modification of dilution rate. Bioresour Technol 220:253–261
Devi MP et al (2012) Heterotrophic cultivation of mixed microalgae for lipid accumulation and wastewater treatment during sequential growth and starvation phases: effect of nutrient supplementation. Renew Energy 43:276–283
Dong QL et al (2006) Metabolic flux analysis of the two astaxanthin-producing microorganisms Haematococcus pluvialis and Phaffia rhodozyma in the pure and mixed cultures. Biotechnol J Healthcare Nutr Technol 1(11):1283–1292
Grobbelaar JU (2013) Mass production of microalgae at optimal photosynthetic rates. In: Photosynthesis. InTechOpen, London, pp 357–371
Guo W et al (2020) A novel porous nickel-foam filled CO_2 absorptive photobioreactor system to promote CO_2 conversion by microalgal biomass. Sci Total Environ 713:136593
Hemalatha M et al (2019) Microalgae-biorefinery with cascading resource recovery design associated to dairy wastewater treatment. Bioresour Technol 284:424–429
Hernández D et al (2014) Biofuels from microalgae: lipid extraction and methane production from the residual biomass in a biorefinery approach. Bioresour Technol 170:370–378
Ho S-H et al (2014) Perspectives on engineering strategies for improving biofuel production from microalgae—a critical review. Biotechnol Adv 32:1448–1459
Huang W et al (2019) The role of water-water cycle in regulating the redox state of photosystem I under fluctuating light. Biochim Biophys Acta Bioenerg 1860:383–390
Ling J et al (2014) Lipid production by a mixed culture of oleaginous yeast and microalga from distillery and domestic mixed wastewater. Bioresour Technol 173:132–139
Jacob-Lopes E et al (2015) Microalgal biorefineries. In: Jacob-Lopes E, Queiroz Zepka L (eds) Biomass production and uses. InTechOpen, Rijeka, pp 81–106
Jokel M et al (2015) Chlamydomonas flavodiiron proteins facilitate acclimation to anoxia during sulfur deprivation. Plant Cell Physiol 56:1598–1607
Jokel M et al (2018) Hunting the main player enabling Chlamydomonas reinhardtii growth under fluctuating light. Plant J 94(5):822–835
Kiran BR, Mohan SV (2021) Photosynthetic transients in Chlorella sorokiniana during phycoremediation of dairy wastewater under distinct light intensities. Bioresour Technol 340:125593
Kodru S, Malavath T, Devadasu E, Nellaepalli S, Stirbet A, Subramanyam R, Govindjee (2015) The slow S to M rise of chlorophyll a fluorescence reflects transition from state 2 to state 1 in the green alga Chlamydomonas reinhardtii. Photosynth Res 125:219–231
Kouzuma A, Watanabe K (2015) Exploring the potential of algae/bacteria interactions. Curr Opin Biotechnol 33:125–129

Kumar S et al (2017) Molecular insights into photosynthesis and carbohydrate metabolism in Jatropha curcas grown under elevated CO_2 using transcriptome sequencing and assembly. Sci Rep 7(1):11066

Lakshmikandan M et al (2021) Co-cultivation of Streptomyces and microalgal cells as an efficient system for biodiesel production and bioflocculation formation. Bioresour Technol 332:125118

Malapascua JR et al (2014) Photosynthesis monitoring to optimize growth of microalgal mass cultures: application of chlorophyll fluorescence techniques. Aquat Biol 22:123–140

Mehrabadi A et al (2017) Effect of CO_2 addition on biomass energy yield in wastewater treatment high rate algal mesocosms. Algal Res 22:93–103

Merchuk J, Camacho G (2010) Bioreactors. Air-lift reactors. In: Flickinger MC (ed) Encyclopedia of industrial biotechnology. Bioprocess, bioseparation, and cell technology. Wiley, Hoboken

Mohan SV et al (2011) Potential of mixed microalgae to harness biodiesel from ecological water-bodies with simultaneous treatment. Bioresour Technol 102:1109–1117

Mohan SV et al (2015) Heterotrophic microalgae cultivation to synergize biodiesel production with waste remediation: progress and perspectives. Bioresour Technol 184:169–178

Nie X et al (2020) Current progress, challenges and perspectives in microalgae-based nutrient removal for aquaculture waste: a comprehensive review. J Clean Prod 277:124209

Park J et al (2011) Wastewater treatment high rate algal ponds for biofuel production. Bioresour Technol 102:35–42

Park KY et al (2013) Anaerobic digestion of microalgal biomass with ultrasonic disintegration. Int Biodeterior Biodegradation 85:598–602

Posadas E et al (2013) Carbon and nutrient removal from centrates and domestic wastewater using algal–bacterial biofilm bioreactors. Bioresour Technol 139:50–58

Posadas E et al (2015) Influence of pH and CO_2 source on the performance of microalgae-based secondary domestic wastewater treatment in outdoors pilot raceways. Chem Eng J 265:239–248

Ramanan R et al (2016) Algae–bacteria interactions: evolution, ecology and emerging applications. Biotechnol Adv 34:14–29

Ranadheer P, Kona R, Sreeharsha RV, Mohan SV (2019) Non-lethal nitrate supplementation enhances photosystem II efficiency in mixotrophic microalgae towards the synthesis of proteins and lipids. Bioresour Technol 283:373–377

Ras M et al (2013) Temperature effect on microalgae: a crucial factor for outdoor production. Rev Environ Sci Biotechnol 12:153–164

Rashid N et al (2019) Co-cultivation of two freshwater microalgae species to improve biomass productivity and biodiesel production. Energy Convers Manag 196:640–648

Reddy KS, Sekhar KM, Sreeharsha RV, Reddy AR (2019) Hydraulic dynamics and photosynthetic performance facilitate rapid screening of field grown mulberry (Morus spp.) genotypes for drought tolerance. Environ Exp Bot 157:320–330

Rivas MO et al (2010) Interactions of Botryococcus braunii cultures with bacterial biofilms. Microb Ecol 60:628–635

Robles Á et al (2020) Microalgae-bacteria consortia in high-rate ponds for treating urban wastewater: elucidating the key state indicators under dynamic conditions. J Environ Manag 261:110244

Rohit M, Venkata Mohan S (2018) Quantum yield and fatty acid profile variations with nutritional mode during microalgae cultivation. Front Bioeng Biotechnol 6:111

Ryu B-G et al (2017) Feasibility of using a microalgal-bacterial consortium for treatment of toxic coke wastewater with concomitant production of microbial lipids. Bioresour Technol 225:58–66

Santos CA et al (2014) Dual-mode cultivation of Chlorella protothecoides applying inter-reactors gas transfer improves microalgae biodiesel production. J Biotechnol 184:74–83

Saravanan A et al (2021) Effective water/wastewater treatment methodologies for toxic pollutants removal: processes and applications towards sustainable development. Chemosphere 280:130595

Sekhar KM et al (2015) Differential responses in photosynthesis, growth and biomass yields in two mulberry genotypes grown under elevated CO_2 atmosphere. J Photochem Photobiol B 151:172–179

Shen Y et al (2017) Municipal wastewater treatment via co-immobilized microalgal-bacterial symbiosis: microorganism growth and nutrients removal. Bioresour Technol 243:905–913

Shen X-F et al (2022) The mechanism of carbon source utilization by microalgae when co-cultivated with photosynthetic bacteria. Bioresour Technol 365:128152

Simionato D et al (2013) Optimization of light use efficiency for biofuel production in algae. Biophys Chem 182:71–78

Sivaramakrishnan R, Incharoensakdi A (2017) Enhancement of total lipid yield by nitrogen, carbon, and iron supplementation in isolated microalgae. J Phycol 53:855–868

Sreeharsha RV, Sekhar KM, Reddy AR (2015) Delayed flowering is associated with lack of photosynthetic acclimation in Pigeon pea (Cajanus cajan L.) grown under elevated CO_2. Plant Sci 231:82–93

Sreeharsha RV, Venkata Mohan S (2021) Symbiotic integration of bioprocesses to design a self-sustainable life supporting ecosystem in a circular economy framework. Bioresour Technol 326:124712. https://doi.org/10.1016/j.biortech.2021.124712

Subhash GV et al (2014) Temperature induced stress influence on biodiesel productivity during mixotrophic microalgae cultivation with wastewater. Bioresour Technol 169:789–793

Sutherland DL et al (2017) Environmental drivers that influence microalgal species in fullscale wastewater treatment high rate algal ponds. Water Res 124:504–512

Tang DYY et al (2020) Potential utilization of bioproducts from microalgae for the quality enhancement of natural products. Bioresour Technol 304:122997

Tikhonov AN (2014) The cytochrome b6f complex at the crossroad of photosynthetic electron transport pathways. Plant Physiol Biochem 81:163–183

Tikkanen M et al (2017) Comparative analysis of mutant plants impaired in the main regulatory mechanisms of photosynthetic light reactions-from biophysical measurements to molecular mechanisms. Plant Physiol Biochem 112:290–301

Uggetti E et al (2018) Start-up of a microalgae-based treatment system within the biorefinery concept: from wastewater to bioproducts. Water Sci Technol 78:114–124

Ullah K et al (2015) Assessing the potential of algal biomass opportunities for bioenergy industry: a review. Fuel 143:414–423

Vadiveloo A, Moheimani N (2018) Effect of continuous and daytime mixing on Nannochloropsis growth in raceway ponds. Algal Res 33:190–196

Villa JA et al (2014) Azotobacter vinelandii siderophore can provide nitrogen to support the culture of the green algae Neochloris oleoabundans and Scenedesmus sp. BA032. FEMS Microbiol Lett 351(1):70–77

Wang H et al (2014) Joint production of biodiesel and bioethanol from filamentous oleaginous microalgae Tribonema sp. Bioresour Technol 172:169–173

Wang Y et al (2016) Perspectives on the feasibility of using microalgae for industrial wastewater treatment. Bioresour Technol 222:485–497

Wimalasekera R (2019) Effect of light intensity on photosynthesis. In: Photosynthesis, productivity and environmental stress. Wiley, Hoboken, pp 65–73

Wrede D et al (2014) Co-cultivation of fungal and microalgal cells as an efficient system for harvesting microalgal cells, lipid production and wastewater treatment. PLoS One 9(11):e113497

Wu Y et al (2013) Isolation, identification and characterization of an electrogenic microalgae strain. PLoS One 8:e73442

Xie Z et al (2021) Co-cultivation of microalga and xylanolytic bacterium by a continuous two-step strategy to enhance algal lipid production. Bioresour Technol 330:124953

Yen H-W et al (2015) The synergistic effects for the co-cultivation of oleaginous yeast-Rhodotorula glutinis and microalgae-Scenedesmus obliquus on the biomass and total lipids accumulation. Bioresour Technol 184:148–152

Yin Z et al (2020) A comprehensive review on cultivation and harvesting of microalgae for biodiesel production: environmental pollution control and future directions. Bioresour Technol 301:122804

Yong JJJY et al (2021) Prospects and development of algal-bacterial biotechnology in environmental management and protection. Biotechnol Adv 47:107684

Zhang K et al (2014) Lipid production and composition of fatty acids in Chlorella vulgaris cultured using different methods: photoautotrophic, heterotrophic, and pure and mixed conditions. Ann Microbiol 64:1239–1246

Zhao P, Yu X, Li J, Tang X, Huang Z (2014) Enhancing lipid productivity by co-cultivation of Chlorella sp. U4341 and Monoraphidium sp. FXY-10. J Biosci Bioeng 118(1):72-77

Zivcak M et al (2015) Repetitive light pulse-induced photoinhibition of photosystem I severely affects CO_2 assimilation and photoprotection in wheat leaves. Photosynth Res 126:449–463

Chapter 5
Novel Approaches in Wastewater Utilization Using Photosynthetic Microbes

Abstract The global industrialization has led to significant changes in ecological dynamics and biodiversity. Population growth, urbanization, and changes in consumption patterns have influenced human interactions with the environment. The release of wastewater into mainstream aquatic channels without proper pretreatment can lead to serious problems, including eutrophication and algal blooms that disturb aquatic life forms. Improving wastewater treatment and management is crucial for protecting human health, ensuring sustainable development, and protecting the environment. The use of microalgae and cyanobacteria for wastewater treatment is a cost-effective and environmentally friendly approach that not only improves water quality but also produces value-added products such as biofuels, food, pharmaceuticals, and cosmetics. Integrating biorefinery techniques with wastewater treatment has several sustainable advantages including cost-effectiveness. Bioelectrochemical systems (BESs) have significantly increased in the past decade, providing efficient and clean production of high-value compounds and fuels using microbes. BESs can be classified into several distinct categories, namely electrogenesis systems, electro-hydrogenesis systems, microbiological desalination processes systems, microbiological electrosynthesis systems, and bioelectrochemical systems for water treatment. Microbial fuel cells (MFCs) are a type of bioreactors that facilitate the conversion of chemical energy from organic substances into electrical energy. This conversion process is achieved by the catalytic process of microbes, which occurs in anaerobic conditions. The current chapter highlights several of these sustainable techniques employed in wastewater treatments with the advent of photosynthetic microbes.

Keywords Wastewater · Bioelectrochemical system · Microbial fuel cells · Electrogenesis · Biorefinery

R. V. Sreeharsha, S. Venkata Mohan, *Microbial Photosynthesis*,
https://doi.org/10.1007/978-981-97-1253-3_5

5.1 Introduction

Global industrialization has revolutionized human life and concomitantly changed the ecological dynamics and biodiversity. Population growth, urbanization, and changes in consumption patterns are influencing human interaction with the environment. When compared to the preindustrial era, the aquatic and terrestrial life forms in the industrial era have taken dynamic shifts to adjust to the changing stoichiometry of water and land. The biogeochemical cycles of nitrogen, phosphorus, and water undergo selective pressures and change the equilibriums due to the admixture of organic and inorganic pollutants (Sreeharsha and Venkata Mohan 2021). Water and land pollutions are the serious issues the world is facing now which need to be addressed sustainably. Clean water and sanitation are one of the new-age sustainable development goals which directly or indirectly fulfil other goals including life below water and life on land, good health, and well-being, and the creation of sustainable cities and communities. As the development of industry and infrastructure is inevitable due to the growing needs of humankind, the development should be coupled with innovation in generating minimal waste components, as well as effectively utilizing the generated waste. Photosynthetic microbes through their ability to fix atmospheric carbon and their efficacy in resource recovery from waste streams play a crucial role in wastewater utilization.

The wastewater can be defined as any aquatic end product generated out of primary usage and consists of high chemical oxygen demand (COD) and other toxic substances which make it difficult for further use. The wastewater has many anthropogenic-related routes of generation including municipal, industrial, dairy, and domestic activities, and wastewater generated from each source is different in its composition and chemical nature. The release of thus generated wastewater into mainstream aquatic channels without any pretreatment can lead to an unusual increase in the nutritional composition of water bodies (Sreeharsha et al. 2022). Subsequently, this creates serious problems including eutrophication and algal blooms which disturb the aquatic life forms. Currently, the world produces about 2.01 billion tonnes of municipal solid waste (MSW) per year which includes waste generated from households, commercial establishments, and institutions. The total municipal solid waste produced in India per year is estimated to be around 62 million tonnes as of 2021. The volume of wastewater generated worldwide varies depending on several factors, including population size, urbanization, industrial activities, and level of economic development. The estimated volume of wastewater generated globally is about 1.8 trillion cubic metres (1.8×10^{12} m^3) per year out of which industrial wastewater constitutes 300 billion cubic metres (300×10^9 m^3) and domestic wastewater is about 1.4 trillion cubic metres (1.4×10^{12} m^3) per year (Economic UNDO, United Nations World Water Development Report 2021). In India, industrial wastewater accounts for about 29.2 billion cubic metres (29.2×10^9 m^3) per year while domestic wastewater accounts for 22.7 billion cubic metres (22.7×10^9 m^3) per year. About 80% of wastewater is not treated and disposed of properly, contributing to the pollution of water bodies and leading to environmental and health hazards.

The wastewater and wastewater sludge contain valuable nutrients, and nutrient recovery is crucial for wastewater discharge and production of value-added compounds. Currently, the popular nutrient recovery methods include active sludge process, chemical precipitation, nitrification-denitrification, adsorption, anaerobic digestion, and other contemporary methods like bioelectrochemical systems (BESs) and osmotic membrane bioreactors (OMBRs) are also found useful (Iorhemen et al. 2019; Nuramkhaan et al. 2019). Various methods of wastewater treatment including chemical, physical, and biological processes are represented in Fig. 5.1. Specifically, the removal of phosphate from the sludge phase can be achieved using wet-chemical and thermochemical treatments, which can then be combined with the following procedures for recovery. It is worth noting that the majority of nutrient recovery technologies are often employed in the liquid phase. Moreover, the biological retrieval of phosphate can be achieved by transferring it from the liquid phase to the sludge phase.

Improving wastewater treatment and management is crucial for protecting human health, ensuring sustainable development, and protecting the environment. The United Nations has set a target to substantially increase the proportion of wastewater treated and safely reused globally by 2030 as part of its Sustainable

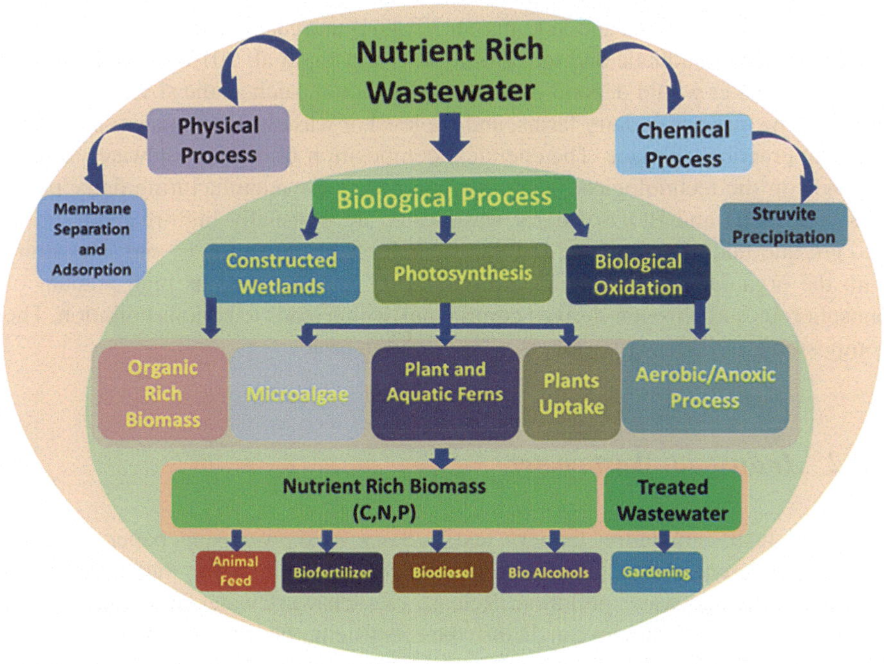

Fig. 5.1 Schematic representation of different methods for wastewater treatment highlighting the biological methods. (Reprinted with permission from Mohan et al. 2016)

Development Goals. The establishment of regulatory frameworks and standards, the promotion of cleaner production practices, and the incentivization of investment in wastewater treatment infrastructure can significantly improve wastewater management. The Indian government has launched several initiatives to improve domestic wastewater treatment and management, including the Swachh Bharat Mission, which aims to improve sanitation infrastructure and promote behavioral change among citizens to reduce open defecation. The government is also promoting the use of decentralized wastewater treatment systems, such as constructed wetlands and biogas plants, to treat and reuse wastewater sustainably.

5.2 Sources and Composition of Different Wastewater

5.2.1 Dairy Wastewater

The dairy industry is consistently growing due to the demand for milk and milk-based products and this in turn increased the production of dairy wastewater. According to the Food and Agriculture Organization of the United Nations, global milk production is increasing at a rate of 2.2% and accounted for 843 million tonnes in 2018. Typically, the dairy sector produces a substantial quantity of waste, ranging from 0.2 to 10 litres per litre of processed milk, necessitating significant management efforts (Ummalyma and Sukumaran 2014). If untreated, dairy wastewater represents a threat to aquatic and terrestrial life (Karadag et al. 2015). The amount of dairy wastewater would depend on a variety of factors such as the size of the dairy industry, the number of dairy farms, and the level of wastewater treatment and management practices in place. The chemical composition of dairy wastewater usually depends on the technological operations employed for manufacturing dairy products. The COD and BOD of dairy wastewater are unusually high reaching up to 120 g/L and majorly constitute D-Lactone, soluble proteins, lipids, and salts along with the organic contents (Kolev Slavov 2017). It also contains high amount of phosphorous and nitrogen-derived compounds which leads to serious pollution. The composition of dairy wastewater is represented in table.

5.2.2 Industrial Wastewater

Diverse industries including tannery, textile, dairy, food processing, paper and pulp, and brewery generate large amount of wastewater which is the major source of environmental and water pollution. Release of such wastewater into water bodies without proper pretreatment can disturb the ecosystem, thus affecting human health. The composition of industrial wastewater can vary widely depending on the type of industrial activity and the specific manufacturing processes used. In general, industrial wastewater contains a variety of organic and inorganic pollutants, including suspended solids such as dirt, silt, and heavy metals such as lead, cadmium, chromium, and mercury, as well as organic compounds including solvents, fuels, lubricants, and other chemicals used in manufacturing processes. Further, they also contain acids, bases, and nutrients such as nitrogen and phosphorus. These elements can clog waterways which reduce water quality and cause corrosion in water treatment systems. They can contribute to eutrophication in water bodies which can be harmful to aquatic life. The adverse effects of industrial wastewater can be curtailed by using measures such as reducing the number of pollutants generated through cleaner production processes, treating wastewater to remove pollutants before discharge, and implementing appropriate disposal methods for residual waste. Regulatory agencies often establish discharge limits for various pollutants in industrial wastewater to ensure compliance with environmental regulations.

5.2.3 Agriculture Wastewater

The agrarian ecosystem including livestock and animal husbandry demands large amount of water consumption and leaves behind contaminated effluents. The composition of agricultural wastewater can vary depending on the type of agricultural activity and the specific management practices used. Broadly, agricultural wastewater is a multifaceted amalgamation of both organic and inorganic substances, encompassing essential elements like nitrogen and phosphorus that can originate from sources such as fertilizers, animal excrement, and other forms of organic material. In addition, agriculture wastewater also contains pesticides and herbicides, sediments due to soil erosion and surface runoff, organic matter including crop residues, animal waste, and other organic debris, and pathogens such as bacteria, viruses, and parasites, which can pose a risk to human and animal health. The generation of wastewater from agriculture sources can be minimized by reducing nutrient inputs, promoting conservation tillage and other soil conservation practices, and implementing appropriate treatment and disposal systems for wastewater. In recent years, there has been an observed increase in the levels of fertilizers, including ammonium and phosphate, in aquatic environments. This can be attributed to the high agricultural output, which has led to unregulated wastewater discharges. Aquatic ecosystems have the potential to undergo eutrophication, a process that can have detrimental effects on both human well-being and the natural environment (Meena et al. 2019; Ye et al. 2019).

5.2.4 *Municipal Wastewater*

Municipal wastewater possesses a diverse combination of organic and inorganic compounds, along with bacteria and pathogens. The composition of such wastewater is contingent upon factors such as the size and geographical location of the municipality, the time of year, and the level of industrial operations within the locality. Typical effluents from municipal wastes contain carbohydrates, proteins, fats, and oils, nutrients such as nitrogen and phosphorus, and inorganic compounds including salts, metals, and trace elements.

5.3 The Concept of Biorefinery—A Multifaceted Strategy

The biorefinery concept encompasses a sustainable approach to biomass conversion, wherein integrated, efficient, and flexible processing techniques are employed to yield a diverse range of goods such as chemicals, food, animal feed, and energy commodities (Sadhukhan et al. 2014). Bioprocess and biorefinery nexus will help in utilizing renewable resources to convert into sustainable fuels and chemicals. The integration of diverse bioprocessing techniques will create an efficient and cost-effective way to valorize biomass. The materials generated through individual bioprocessing techniques such as fermentation and biotransformation can be further converted into other products through biorefinery technologies. Organizations including the National Renewable Energy Laboratory (NREL), the U.S. Department of Energy (DOE), and the Bioprocess and Biorefinery Nexus Consortium (BBNC) have studied various aspects of the bioprocess and biorefinery integration and developed new technologies. Microalgae and cyanobacteria are proved to be an efficient resource for biorefinery applications in renewable energy, fuels, and chemicals (Sreeharsha et al. 2023; Jin et al. 2015). They help in reducing the dependence on petroleum products and lower emissions from fossil fuels. The recent advancements in metabolic engineering technologies have improved the efficiency of the microbial biorefineries making them robust cell factories. High yield per square metre, making use of unusable terrains, using a wide range of water sources (fresh, brackish, saline, marine, wastewater), production of biofuels and other valuable by-products, and recycling CO_2 from atmosphere are the positive aspects of microbial biorefinery. Nevertheless, the safety and quality of the products, economic viability, and ethical considerations concerning the environment and human health need to be taken care in these biorefineries.

5.3.1 *Microalgae Biorefinery for Wastewater Treatment and Valorization*

Nutrient recovery from wastewater is an important and growing field of research to reduce pollution and nutrient loading in aquatic ecosystems. Chemical methods of wastewater treatment including the use of oxidation treatments with ozone or hydrogen peroxide convert nitrogen and phosphorus into more readily recoverable forms. Such treatments are highly efficient but can be costly and may require additional pretreatment. Such traditional methods of wastewater treatment are expensive and non-environment friendly because of this, scientists have started searching for eco-friendly biological treatment methods. The use of microalgae for wastewater treatment could be a cost-effective and environmentally friendly approach, wherein it improves water quality besides producing value-added products which are alternatives to traditional compounds. Microalgae are flexible enough to grow in varied habitats including ponds, tanks, or wastewater lagoons, thus integrating with wastewater biorefinery. The recent developments in large-scale culture systems, biomass harvesting, and processing techniques benefited the use of microalgae for wastewater treatment (Kumari et al. 2022). The nitrogen, phosphorus, and other nutrients present in wastewater can act as resources for microalgae growth and biomass production through photosynthesis. Wastewaters are usually abundant in trace metals including Fe, Cu, Mn, and Zn, as well as important nutrients such as phosphate (PO_4), ammonia (NH_3), and other minerals. Most freshwater microalgal strains can produce superior biomass concentrations by rapidly degrading these macro- and micro-nutrients. Additionally, the COD can be removed thereby reducing the toxicity of the wastewater effluents (Chandra et al. 2021; Khaldi et al. 2017). The utilization of nutrients present in wastewater in a way eliminates the use of overpriced fertilizers for large-scale cultivation systems of microalgae. Algal-based wastewater treatment technologies are becoming attractive owing to their potential in less energy consumption compared to traditional wastewater treatment systems (0.2 kW/h/m^3, compared to 2 kW/h/m^3, respectively) making it more energy efficient (Craggs et al. 2012).

In a biorefinery approach, microalgae could present multiple benefits of producing a variety of products such as biofuels, food supplements, pharmaceuticals, cosmetics, antibacterial, antiviral, antitumor/anticancer, and antihistamine compounds, thus providing environment-friendly solutions to energy and resource demands. Multiple inputs from diverse industrial wastes generate multiple outputs, thus maximizing the resource recovery and reducing the environmental impact. In recent times, the microalgae biorefinery assimilated innovative techniques to growth and biomass production and typically involves a series of processes including harvesting, extraction, and conversion, to produce the desired products. Life cycle assessment and economic analysis of microalgae biorefinery suggest that the majority of newly isolated microalgal strains can produce cost-effective bioproducts with the benefits of reducing greenhouse gases and pollutants (Okeke et al. 2022).

In the last two decades, several microalgae species including *Chlorella pyrenoidosa, Anabaena ambigua, Scenedesmus abundans, Chlorella vulgaris, Chlamydomonas polypyrenoideum* and *Acutodesmus dimorphus, Chlorella minutissima, Nostoc muscorum*, and *Spirulina* sp., have been successfully cultivated in dairy wastewater (Brar et al. 2019; Choi 2016; Chokshi et al. 2016; Kothari et al. 2013; Qin et al. 2014; Ranadheer et al. 2019). *Chlorella* sp. shows promising results in removing phosphorus and nitrogen form piggery wastewater, thus making it a reliable resource for agriculture wastewater treatment (Wang et al. 2016). The utilization of piggery wastewater as a growth medium for *C. zofingiensis* achieved 3 g/L biomass which is very significant. The piggery wastewaters which predominantly show the presence of propionic acid, acetic acid, and butyric acids facilitate mixotrophic growth of microalgae. This considerably increases biomass and lipid production, thus valorizing the pig farming industry.

In general, the wastewater generated from the source is highly concentrated that possesses higher COD and needs to be diluted before microalgae treatment. Researchers need to continue to explore the economic and environmental benefits of this technology, as well as to develop innovative techniques to maximize its efficiency (Sivaramakrishnan and Incharoensakdi 2019). Lipids extracted from algae using wastewater as a feedstock are listed in a tabular form (Table 5.1).

The biorefinery approach can be used to generate a variety of products by using different biochemical and thermochemical treatments, for instance, alcoholic

Table 5.1 Summary of microalgae grown in wastewater to produce lipids

Algal species	Wastewater used	Lipid content
Scenedesmus sp.	Fermented swine wastewater	9 mg/g
Botryococcus braunii	Secondary treated sewage	17%
Chlorella pyrenoidosa	Piggery wastewater	6.3 mg/L/day
Chlorella pyrenoidosa FACHB-9	Soybean process wastewater	37%
Chlorococcum sp. *RAP13*	Dairy effluent	42%
Chlamydomonas Reinhardtii	Artificial wastewater	25.2%
Auxenochlorella protothecoides UMN 280	Municipal wastewater	28%
Chlorella vulgaris	Agro-industrial co-products, ethanol thin stillage, and soy whey	24%
Chlorella vulgaris	Food industrial wastewater	25.4%
Chlorella vulgaris	Brewery wastewater	26.7%
Oedogonium intermedium	Treated municipal wastewater	10%
Chlorella zofingiensis	Treated municipal wastewater/biogas slurry	8%
Algal consortium	Anaerobic digested piggery effluent	32.7%

fermentation of deoiled algal biomass cake following lipid extraction, the production of biogas, and the combustion of this gas to yield bioelectricity. Wastewater treatment in a biorefinery framework utilize the residual algal biomass effectively, and promote sustainable bioprocess with better economics. Upstream and downstream processing are the two primary steps that are regarded as crucial in the microalgal biorefinery. The types of microalgae strains, nutrition supplies, and solar illumination are all considered upstream processing factors (Daneshvar et al. 2021; Ehimen et al. 2011; Vanthoor-Koopmans et al. 2013).

5.3.2 *Bacterial Biorefinery*

Bacteria are diverse group of organisms with multiple industrial applications and can be attractive resource for biorefinery framework. In particular, photosynthetic bacteria are useful in producing value-added compounds, bioelectricity, fuels, and chemicals, using the nutrients present in the wastewater effluents. The ability of bacteria to produce bioelectricity through multidimensional mechanisms places certain species of bacteria in forefront in biorefinery applications. The global energy requirements are rising consistently wherein the average energy consumption is projected to 23 TW by 2050 from the current 13 TW (Yang et al. 2019). Due to the disadvantages of fossil fuels, there is an increasing demand for alternative energy sources which help in sustainability and reduction in energy imports (Resch et al. 2008). The use of bioelectrochemical systems (BESs), a burgeoning sustainable technology for the concurrent generation of energy and wastewater treatment, has experienced a substantial upsurge during the last decade (Mohan et al. 2010). Furthermore, BESs provide distinct prospects for the effective and environmentally friendly synthesis of valuable chemicals and fuels through the utilization of microorganisms (ElMekawy et al. 2015). The redox potentials of the oxidation-reduction processes occurring at the anode and cathode inside electrochemical systems generate a potential difference that serves as the driving force for the spontaneous movement of electrons from a region of lower potential to a region of higher potential (Patil et al. 2015). Electric current is the unit of measurement for this flow of electrons across an external circuit. When bacteria or enzymes are involved in such oxido-reduction reactions to generate redox potential, it is referred to a biological electrochemical system (BES) (Harnisch and Schröder 2010). Extensive study has been conducted on a global scale in several fields such as microbiological science, the electrochemistry, bioelectrochemistry, the field of biotechnology environmental science, and materials science, owing to the interdisciplinary character of BES.

Bioelectricity is commonly generated by the oxidation of organic matter at the anode, which occurs at a low redox potential. Simultaneously, the reduction of oxygen takes place at the cathode, operating at a high redox potential. In the context of microbial electrolysis cells, the synthesis of significant chemicals is facilitated by the manipulation of cathode potentials by the application of an external voltage. Microbial electrosynthesis (MES) refers to a BES wherein carbon dioxide (CO_2) or organic molecules undergoes cathodic conversion, resulting in the production of organic chemicals with enhanced efficiency (ElMekawy et al. 2014; Rabaey and Rozendal 2010). BESs have seen further advancements, involving several applications including microbial desalination cells (MDCs) for water and sediment desalination, as well as plant microbial fuel cells that include the symbiotic pairing of bacteria with plant roots to generate energy (Cao et al. 2009; Chiranjeevi et al. 2012; Ehimen et al. 2011). Additionally, BESs are recognized as effective bioreactors for the bioelectrochemical treatment (BET) or microbial electroremediation of refractory contaminants and toxic wastewater (Mohanakrishna et al. 2015). BESs may be broadly classified into multiple categories based on their intended outcomes. These categories include electrogenesis systems, electrohydrogenesis systems, microbiological desalination systems, microbiological electrosynthesis systems, and bioelectrochemical systems for water treatment.

5.3.3 Microbial Fuel Cell

Microbial fuel cells (MFCs) are "bioreactors that convert the chemical energy of the organic compounds into electrical energy by the catalytic activity of microorganisms under anaerobic environments" (Du et al. 2007). A conventional two-chambered MFC with an anode and a cathode chamber is divided by a selectively permeable membrane. The anode of an MFC easily satisfies the microorganisms' requirement for a compatible electron acceptor to deposit electrons in the absence of a more compatible acceptor (Stams et al. 2006). To generate useful energy, the anode's electrons are directed across a resistor, an external load. The electron transport process concludes at the cathode, where there is a terminal electron acceptor. The fundamental component of a microbial fuel cell (MFC) may be described as a "quasi-engineered" electron transport chain, which emulates the respiratory chain found in bacteria. These microbe-catalyzed electrochemical devices are seen as a possible source of sustainable and green energy as well as a wastewater management solution. The primary focus of research in microbial electrochemical technologies has been on four key variables that are crucial for achieving practical feasibility. These factors encompass the reduction of electrochemical losses, the improvement of performance efficiency, the reduction of operational expenses, and the development of scalable systems for real-world applications.

In an MFC, microbes are essential because they catalyze the release of electrons from organic substrates' energy-rich bonds when the environment is anoxic. Various types of pure substrates and diverse wastewater samples have been utilized as carbon sources to support the growth of microorganisms in microbial fuel cells (MFCs) (Pandey et al. 2016; Pant et al. 2010). During the oxidation process, the electrons are discharged and subsequently traverse through several microbial electron transport chains prior to their contribution to the anode of the MFC. These chains consist of consecutively ordered conductive complexes of protein, cytochromes, nanowires, and redox proteins (Fredrickson et al. 2008; Kracke et al. 2015; da Costa 2018).

Either indirect electron transfer, which is mediated by naturally occurring or chemically inserted redox shuttles, or direct extracellular electron transfer is possible between bacteria and the electrode (Kumar et al. 2017; Lovley 2012; Martinez and Alvarez 2018).

Compared to pure bacterial cultures, mixed consortia of electrogenic and electrotrophic microorganisms have been shown to more efficiently contribute to the production of current in MFCs (Logan et al. 2019). This discrepancy could be explained by the development of electrochemically active biofilms, which emerge from the synergistic interactions between syntrophic microbial species (Kiely et al. 2011; Reguera 2018; Kiran and Patil 2019). By applying synthetic biology, selectively manipulating growth conditions, and implementing engineering methods, electroactive biofilm growth and performance can be improved (Angelaalincy et al. 2018; Chiranjeevi and Patil 2020; Doyle and Marsili 2018; Glaven 2019). Additionally, to gain insights from a structural and functional perspective, communities of microbial consortia have been surveyed and described utilizing "omics" technologies, flowcytometric techniques, and computational tools (Franzosa et al. 2015; Koch et al. 2019; Kouzuma et al. 2018). MFCs employ an electrode that is solid as an acceptor of electrons in order to generate electrical current through the process of microbial oxidation of organic substances (Rabaey et al. 2011).

The oxidation of organic materials by microbes on the anode surface of MFCs results in the generation of electrons. These electrons are then transferred to the cathode compartment through an external circuit, which also contains an external load. Ions moving through a membrane or other ion-permeable media as electricity is generated ensures electroneutrality. M.C. Potter originally showcased one of these devices in 1911 that achieved a maximum voltage of 0.3–0.5 V.

The electrocatalytic component of the BES, which produces energy, is composed of electroactive bacterial biofilms that have grown on the anode. The presence of exoelectrogens, which are microorganisms capable of generating electricity, on the anode plays a crucial role in facilitating the transfer of electrons from the reduced substrate to the anode. This direct relationship between exoelectrogen presence and bioelectricity generation is observed in MFCs. The oxidation of complex organic compounds in wastewater necessitates the presence of diverse microbial communities, whereas the ability to create electric current is exhibited by different bacterial species. The *Shewanella* and *Geobacter* species have been widely investigated as electron-transferring microorganisms due to their recognized capabilities in this regard. In contrast to microbial fuel cells (MFCs) utilizing cultures that are pure in the bio-anodes, a greater occurrence of constant and higher currents has been found in those systems. In MFCs, high power densities frequently indicated the presence of Geobacteraceae in the bio-anode community (Kiely et al. 2011). The layered-corrugated carbon anode with a high surface area yielded a 390 A m^2 current density, which is the highest current density in MFC to date (Chen et al. 2012). Electrolyte chemistry, anode potential, and substrate type all have an impact on microbial activity and electron transport. The performance of microbial electrochemical systems is improved by higher electrolyte conductivity, but this conductivity should not exceed the threshold that bacteria can tolerate (Sharma et al. 2014).

Nevertheless, there are reports indicating that the employment of halophilic bacteria in BESs has demonstrated the ability to generate current densities of up to 85 A m^2. These bio-anodes are derived from a salt-marsh inoculum and cultivated in an electrolyte that is 1.5 times more saline than seawater (Rousseau et al. 2013). MFC performance is also influenced by the structure and material of the electrode. Roughened-surface graphite electrodes have been proven to generate better power densities than flat graphite electrodes. Comparable findings have been achieved by the augmentation of the surface area of MFC anodes via the utilization of porous substances, including graphite brush, carbon felt, and carbon nanotubes (Logan et al. 2019; Wilberforce et al. 2021).

An increased surface area of electrode facilitates enhanced adhesion of bacteria and transfer of electrons, leading to a discernible augmentation in current densities. The direct electron transfer phenomena between microorganisms and metal electrodes have been successfully created using dimensionally stable anodes (DSA) composed of titanium over iridium, tantalum oxide, metals such as stainless steel, and platinum electrodes (Pocaznoi et al. 2012). With the exception of platinum, the current densities seen using porous carbon anodes frequently surpass those documented with metals. The performance of bioelectrodes was significantly improved when three-dimensional electrodes with tailored microstructures were used instead of planar electrodes (Sharma et al. 2014). A significant increase in current densities was demonstrated by macrostructure alteration in the anode, with a multi-layered corrugated carbon arrangement. Low internal resistances enable it to obtain larger current and power densities, particularly after flow-through electrodes are used to overcome mass transport restrictions (Sleutels et al. 2012). Catalysts such as MnO_2-coated air cathodes and membranes and cloth electrodes assemblies employed as cathodes in MFCs have demonstrated higher power densities. The internal resistance, which in turn controls the system's overall performance, is influenced by the materials used in the MFC's construction, design, and configuration.

5.4 Bioelectrical Systems

5.4.1 Bioelectrochemical Treatment

Microbial electrochemical technologies enable the use of the generated power for a variety of purposes beyond just bioenergy, such as the treatment of wastewater and the manufacturing of biobased goods (METs). The organic and inorganic compounds in the wastewater were utilized as a potential feedstock for microbially catalyzed systems, in which a variety of reactions, including biochemical, physical, physicochemical, electrochemical, oxidation, etc. occur. These reactions have collectively been referred to as "bioelectrochemical reactions," and the procedure is known as "bioelectrochemical treatment (BET)" (Mohan et al. 2010, 2019; Velvizhi and Mohan 2011). The process of BET involves the use of electrochemically active

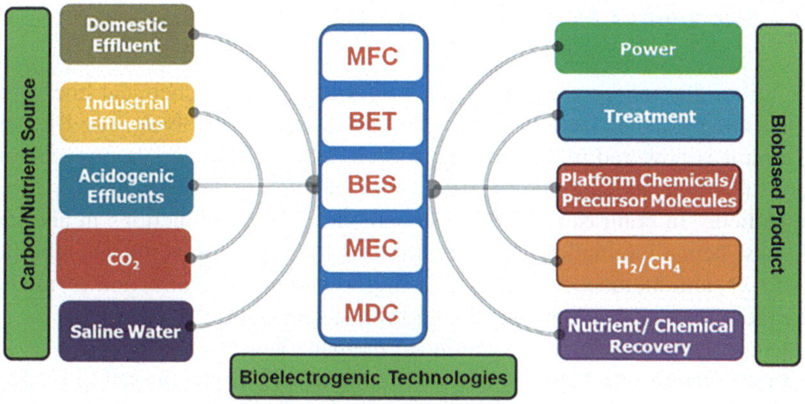

Fig. 5.2 Schematic representation of bioelectrochemical systems used for wastewater treatment using photosynthetic microbes. (Reprinted with permission from Mohan et al. 2016)

bacteria (EAB) as a catalyst to facilitate the oxidation of organic substrates present in wastewater, accomplished through a sequence of redox reactions (Mohan et al. 2008; Rozendal et al. 2006). The approach of recovering energy from organic substrates in the form of electricity via using electrochemically active microorganisms as catalysts in fuel cells is now redirected toward the treatment of pollutants (Mohan et al. 2008). To address complicated contaminants, researchers began to reorient the idea of fermentative metabolism toward bioelectrochemical remediation. Anaerobic fermentation, electrolytic dissociation, and electrochemical oxidation are only a slew of the various unit activities that make up BET's integrated wastewater treatment system, which operates simultaneously and synergistically (Fig. 5.2).

The key step in this process is anaerobic fermentation, during which the metabolism of the substrate occurs. The reducing equivalents obtained during this process were deposited on solid electrodes, where they began to start electrochemical oxidation that ultimately results in the electrolytic dissociation of the pollutants. The BET system is an efficient method for removing inorganic pollutants, including salt, employing in situ biopotential created by wastewater degradation that is negligible in traditional anaerobic processes (Mohan et al. 2019; Nancharaiah et al. 2019; Sravan et al. 2020). With regard to COD, color, nitrates, sulfates, certain hydrocarbons, metals, etc., the BET system's numerous procedures minimize pollution (Nancharaiah et al. 2016; Velvizhi and Mohan 2012). Anodic oxidation (direct anodic oxidation [DAO] and indirect anodic oxidation) and cathodic reduction processes are the key mechanisms in the BET system for the treatment of contaminants. BET is governed by several key factors that control the effectiveness of their treatment, including physical, operational, and biological factors.

5.4.2 *Microbial Electrolysis Cells (MECs)*

MECs utilize the inherent capability of bacteria to convert chemical energy into electrical energy, hence facilitating the process of water electrolysis. The BES electrical circuit is powered by an external source, which facilitates the movement of electrons from the anode to the cathode, hence facilitating the creation of hydrogen at the cathode. In contrast to MFCs, the cathode of MECs functions in anaerobic conditions, hence facilitating the generation of hydrogen. Nevertheless, the presence of CO_2 and methanogens in MECs might lead to the production of methane due to the anoxic conditions and the significant generation of hydrogen. Several approaches may be employed to minimize methane generation, including aerating the cathode chamber in between batches, reducing pH levels, operating at shorter retention durations, subjecting the inoculum to heat shock, and adding substances that inhibit the development of methanogens (Logan et al. 2019). When comparing MECs to MFCs, higher electric currents are frequently found and this is because the greater applied voltage assists in overcoming the cathode restrictions in MEC (Sleutels et al. 2012). An additional separate MFC can be used as a power source to supply the energy needed for MEC operation (Wang et al. 2011). The use of fermentation effluent as a substrate was investigated, and a research study documented a hydrogen production rate of 0.24 m^3—H2 m^3 dL, along with an energy recovery efficiency of 23%. However, efficiently connected multiple MFCs are necessary for the provision of a specified voltage that enables maximum hydrogen synthesis in MEC. In a MEC including a bio-anode and a biocathode, the use of expensive catalysts such as platinum is rendered unnecessary. Moreover, the preferential colonization of microorganisms on the carbon cathode leads to a reduction in the time required for system initialization, while simultaneously generating current densities that are equivalent to those achieved at the bio-anode (Jeremiasse et al. 2010). Additionally, the hydrogen generated in MECs can fuel the metabolic synthesis of other compounds (Rabaey and Rozendal 2010). The cathode commonly experiences reduction processes, which encompass the reduction of protons to hydrogen, the reduction of oxygen to hydrogen peroxide (H_2O_2), the reduction of carbon dioxide to methane (CH_4), and the reduction of acetate (Rozendal et al. 2009).

5.4.3 *Microbial Electrosynthesis (MES)*

A cutting-edge approach to BES called MES, also known as bio-electrosynthesis, harnesses the reducing power produced by anodic oxidation to synthesize value-added compounds at the cathode. To create value-added products, cathodic biocatalysts (with associated cathodic biofilms) lower the amount of terminal electron acceptor that is available (Mohanakrishna et al. 2015). The bio-electrosynthesis process can exhibit specificity based on factors such as the biocatalyst involved in catalyzing the redox process, its final electron acceptor, the electrochemically active redox intermediaries, or the suitable reducing equivalents. The essential elements of microbial electrosynthesis are biocathodes, which are used by the electrode-oxidizing bacteria to produce reduced value-added products like acetate, ethanol, and butyrate (Bajracharya et al. 2015). MES involves the use of microorganisms as biocatalysts to facilitate the manufacture of chemical compounds within an electrochemical cell. This process relies on the reduction of CO_2 through the application of electricity, as well as the oxidation and reduction of several other organic feedstocks. MESs may be characterized as microbial catalysts employed for the electrochemical reduction of carbon dioxide (CO_2) into multi-carbon organic molecules. This process relies on the utilization of electrical current at the cathode, serving as a proof of concept. As a result, MES is being considered as a different tactic to capture electrical energy in organic products' covalent chemical interactions.

5.4.4 Enzymatic Fuel Cells (EFCs)

Enzymatic fuel cells (EFCs) employ specific enzymes such as glucose oxidase, immobilized on the electrode surface or suspended in the electrolyte to facilitate the oxidation of fuel and targeted processes required for diverse applications. The extensive use of glucose oxidase in various devices such as pacemakers, indicator lights, mini actuators, micro pumps, and other glucose electron oxidizing anode-based systems may be attributed to its remarkable selectivity and thermostability (Ieropoulos et al. 2008). Based on each enzyme's unique redox activity, different enzymes are applied to anodes and cathodes. The use of highly selective enzymes at the anode and cathode obviates the necessity for a membrane to divide the compartments. Nevertheless, it is commonly seen that enzyme catalysts exhibit incomplete oxidation of the fuel, leading to the production of heat due to concurrent side reactions. This thermal generation may have adverse effects on the enzymatic activity (Minteer et al. 2007). The use of micelle polymers for encapsulation purposes has demonstrated enhanced longevity of enzyme-based systems by providing pH management through buffering capacity and protecting against enzymatic degradation via the creation of a hydrophobic environment. The use of Osmium (Os)-polymer-based electrodes integrated with flavin adenine dinucleotide-dependent glucose dehydrogenases (FADGDH) has exhibited remarkable glucose sensitivity and enhanced current densities. Consequently, these electrodes are deemed suitable for application as bio-anodes in glucose-based enzymatic fuel cells (EFCs). The pyrroquinolinequinone-dependent alcohol dehydrogenase (PQQ-ADH), an enzyme derived from acetic acid bacteria, has been documented for its capacity to engage in quasi-reversible electrochemistry on a carbon surface. This characteristic has led to its application in the development of EFCs and biosensors (Yakushi and Matsushita 2010). According to reports, this enzyme has strong electron transfer kinetics and may use a variety of redox mediators. EFCs are suitable for independently powering wireless sensor applications like video monitoring because of their higher energy densities combined with lower power outputs. They are also suitable for the development of other portable power devices. The anodic catalyst employed in the study was FAD-dependent fructose dehydrogenase derived from Gluconobacter. It shown notable characteristics such as high turnover rates, good stability, and remarkable selectivity. Conversely, the cathode utilized in the investigation was copper-containing laccase sourced from mushrooms. Similar to the anodic catalyst, the copper-containing laccase exhibited high turnover rates, good stability, and excellent selectivity (Kamitaka et al. 2007). To improve the efficiency of the system as a whole and the substrate conversion, multienzyme-based EFCs are typically favored over single enzyme systems. To assist complex mechanisms, several studies are now concentrating on cascades of enzymes in FECs and 3-D electrode topologies that offer enough surface area for enzymatic reactions. The high expense associated with EFCs persists as a result of employing refined enzymes possessing significant catalytic capabilities.

5.4.5 *Microbial Solar Cells*

Microbial solar cells (MSCs) employ higher plants or photoautotrophic microorganisms to harness solar energy, afterwards utilized by electroactive bacteria to facilitate processes involving electrodes. In these reactions, molecules including hydrogen, methane, ethanol, hydrogen peroxide, etc. are produced, as well as electric current. The production of organic chemicals is predominantly achieved through the process of photosynthesis, after which they are then delivered into the anode compartment. Within this compartment, electroactive microorganisms facilitate the oxidation of these organic chemicals, resulting in the generation of electrons. Subsequently, the aforementioned electrons are transferred to the cathodic region, where the process of oxygen reduction leads to the formation of water (ElMekawy et al. 2014). *Synechocystis*, chlorophyta, and other electroactive bacteria are found in the phototrophic biofilms that have formed on the anode. The use of algal biomass produced by Microcystis aeruginosa and Chlorella vulgaris was employed as a substrate within MFCs, in conjunction with lake water. This approach enabled the simultaneous generation of electrical power and treatment of water, within a system designed to address algal blooms in lakes, utilizing ideas derived from MES (Sabourin-Provost and Hallenbeck 2009). After being treated in an anaerobic digester, the algal biomass from photobioreactors can be used as fuel for fuel cells (De Schamphelaire and Verstraete 2009). Chlorella was used as the anode in an MSC photobioreactor, which achieved a light-to-electricity conversion efficiency of 0.04% and an average power density of 14 mW m^2 (Strik et al. 2011). It has been reported that Synechocystis PCC6803 demonstrates a current density of 539 mA m^2 when exposed to elevated light intensity of 10,000 lux. Additionally, this organism has been shown to effectively sequester CO_2, at a rate of 625 mmol CO_2 m^3. This is without a mediator or carbon source in the anolyte. Through the optimization of the photobioreactor, enhanced chemical energy transfer from the algae to the bio-anode, increased surface electrode area, and enriched electroactive biofilm, performance can be improved. The advantages of this system are based on the simultaneous capture of CO_2 and the direct conversion of light to energy.

5.4.6 Plant Microbial Fuel Cells (PMFCs)

Plant microbial fuel cells (PMFCs) are a type of fuel cell that incorporates the root system of a living plant into the anode of a microbial fuel cell (MFC). This integration allows for the utilization of rhizodeposits, which are organic waste products released by the plant roots into the surrounding soil, as a substrate for electroactive bacteria. These bacteria are capable of generating energy through their metabolic processes inside the PMFC system. Thus, in a broader sense, PMFCs harness solar radiation by efficiently and cleanly converting it into green electricity. The primary constituents of rhizodeposits originating from roots consist of exudates, which include sugars and organic acids, as well as secretions including polymeric carbohydrates and enzymes, along with lysates encompassing deceased cell components, and gases. The selection of a plant species for PMFC implementation is of utmost importance, as it plays a direct role in regulating the quantity of rhizodeposits available for the generation of bioelectricity. *Spartina anglica* rice plants (De Schamphelaire et al. 2010), and reed manna grass have all been used in different research. With a maximum power generation of 163 mW m^2, according to reports, Pennisetum setaceum-derived PMFCs have demonstrated the highest level of sustainability in power generation observed to date (Chiranjeevi et al. 2012). The maximum long-term power production of 240 mW m^2 was achieved by *S. anglica* PMFCs with integrated oxygen-lowering biocathodes over a period of 2 weeks (Wetser et al. 2015). The potential of phototrophic biofilms in combination with PMFCs and MSCs as novel and promising energy-harvesting technology is widely recognized. Furthermore, it has been determined that the production of bioelectricity, in contrast to the process of plant photosynthesis, exhibits a direct relationship with the overall quantity of rhizodeposits accessible for oxidation at the anode (Timmers et al. 2012).

5.4.7 Microbial Desalination Cells (MDCs)

Microbial desalination cells (MDCs) use the electrochemical potential gradient established between the anode and cathode through the utilization of microbial fuel cell (MFC) technology for the purpose of conducting on-site desalination. Between the anode and cathode compartments, MDCs have an extra intermediate compartment for desalinating water. The centre compartment is partitioned by an anion exchange membrane (AEM) and a membrane that exchanges cations (CEM), oriented toward the anode and cathode, respectively. To maintain charge neutrality, anions (such as SO_4 Cl), which are present in the desalination compartment, migrate to the anode, and cations (such as K, Na), which are present in the middle chamber solution, are transferred to the cathode (Luo et al. 2011). This results in the desalination of the middle chamber solution. The conductivity of the anolyte and catholyte in MDC is increased by the migration of ions from the middle chamber's saline water toward the anode and cathode. As a result of increased conductivity and mass transfer, electrical power production in MDC has improved. Nevertheless, the salinity of the water may have a detrimental effect on both the anodic/cathodic bacteria and the quality of the effluent water in microbial desalination cells (MDCs).

5.5 Conclusion

Wastewater, generated from various sources, poses significant challenges, including eutrophication and algal blooms, which disrupt aquatic life. About 80% of wastewater is not treated and disposed of properly, contributing to water body pollution and environmental and health hazards. Nutrient recovery is crucial for wastewater discharge and production of value-added compounds. Utilizing eco-friendly biological treatment methods, such as photosynthetic microbes for wastewater treatment, is a sustainable solution when compared to the physical and chemical methods. They can improve water quality and produce value-added products, in a biorefinery approach making them an attractive alternative to traditional compounds. Algal-based wastewater treatment technologies are becoming more energy efficient, as they can grow in various habitats and can produce superior biomass concentrations through photosynthesis. The biorefinery approach can generate various products using different biochemical and thermochemical treatments. Integrating the biorefinery and wastewater treatment can effectively utilize algal biomass, reduce residual waste, and promote sustainable economics. Alternatively, bioelectrochemical treatment (BET) is a promising technology for wastewater treatment and the production of biobased goods.

References

Angelaalincy MJ et al (2018) Biofilm engineering approaches for improving the performance of microbial fuel cells and bioelectrochemical systems. Front Energy Res 6:63

Bajracharya S, ter Heijne A, Benetton XD, Vanbroekhoven K, Buisman CJ, Strik DP, Pant D (2015) Carbon dioxide reduction by mixed and pure cultures in microbial electrosynthesis using an assembly of graphite felt and stainless steel as a cathode. Bioresour Technol 195:14–24

Brar A et al (2019) Comparative appraisal of biomass production, remediation, and bioenergy generation potential of microalgae in dairy wastewater. Front Microbiol 10:678

Cao X et al (2009) A completely anoxic microbial fuel cell using a photo-biocathode for cathodic carbon dioxide reduction. Energy Environ Sci 2:498–501

Chandra R et al (2021) An approach for dairy wastewater remediation using mixture of microalgae and biodiesel production for sustainable transportation. J Environ Manag 297:113210
Chen S et al (2012) Stainless steel mesh supported nitrogen-doped carbon nanofibers for binder-free cathode in microbial fuel cells. Biosens Bioelectron 34:282–285
Chiranjeevi P, Patil SA (2020) Strategies for improving the electroactivity and specific metabolic functionality of microorganisms for various microbial electrochemical technologies. Biotechnol Adv 39:107468
Chiranjeevi P et al (2012) Rhizosphere mediated electrogenesis with the function of anode placement for harnessing bioenergy through CO_2 sequestration. Bioresour Technol 124:364–370
Choi H-J (2016) Dairy wastewater treatment using microalgae for potential biodiesel application. Environ Eng Res 21:393–400
Chokshi K et al (2016) Microalgal biomass generation by phycoremediation of dairy industry wastewater: an integrated approach towards sustainable biofuel production. Bioresour Technol 221:455–460
Craggs R, Sutherland D, Campbell H (2012) Hectare-scale demonstration of high rate algal ponds for enhanced wastewater treatment and biofuel production. J Appl Phycol 24:329–337
da Costa C (2018) Bioelectricity production from microalgae-microbial fuel cell technology (MMFC). MATEC Web of Conferences. EDP Sciences
Daneshvar E et al (2021) Insights into upstream processing of microalgae: a review. Bioresour Technol 329:124870
De Schamphelaire L, Verstraete W (2009) Revival of the biological sunlight-to-biogas energy conversion system. Biotechnol Bioeng 103(2):296–304
De Schamphelaire L, Cabezas A, Marzorati M, Friedrich MW, Boon N, Verstraete W (2010) Microbial community analysis of anodes from sediment microbial fuel cells powered by rhizodeposits of living rice plants. Appl Environ Microbiol 76(6):2002–2008
Doyle LE, Marsili E (2018) Weak electricigens: a new avenue for bioelectrochemical research. Bioresour Technol 258:354–364
Du Z et al (2007) A state of the art review on microbial fuel cells: a promising technology for wastewater treatment and bioenergy. Biotechnol Adv 25:464–482
Economic, U.N.D.o (2021) World social report 2021: reconsidering rural development. United Nations
Ehimen E et al (2011) Anaerobic digestion of microalgae residues resulting from the biodiesel production process. Appl Energy 88:3454–3463
ElMekawy A et al (2014) The near-future integration of microbial desalination cells with reverse osmosis technology. Energy Environ Sci 7:3921–3933
ElMekawy A et al (2015) Food and agricultural wastes as substrates for bioelectrochemical system (BES): the synchronized recovery of sustainable energy and waste treatment. Food Res Int 73:213–225
Franzosa EA et al (2015) Sequencing and beyond: integrating molecular 'omics' for microbial community profiling. Nat Rev Microbiol 13:360–372
Fredrickson JK et al (2008) Towards environmental systems biology of Shewanella. Nat Rev Microbiol 6:592–603
Glaven SM (2019) Bioelectrochemical systems and synthetic biology: more power, more products. Microb Biotechnol 12:819
Harnisch F, Schröder U (2010) From MFC to MXC: chemical and biological cathodes and their potential for microbial bioelectrochemical systems. Chem Soc Rev 39:4433–4448
Ieropoulos I, Greenman J, Melhuish C (2008) Microbial fuel cells based on carbon veil electrodes: stack configuration and scalability. Int J Energy Res 32(13):1228–1240
Iorhemen OT et al (2019) Aerobic granular sludge membrane bioreactor (AGMBR): extracellular polymeric substances (EPS) analysis. Water Res 156:305–314
Jeremiasse AW, Hamelers HV, Buisman CJ (2010) Microbial electrolysis cell with a microbial biocathode. Bioelectrochemistry 78(1):39–43

Jin M et al (2015) Microbial lipid-based lignocellulosic biorefinery: feasibility and challenges. Trends Biotechnol 33:43–54
Kamitaka Y, Tsujimura S, Setoyama N, Kajino T, Kano K (2007) Fructose/dioxygen biofuel cell based on direct electron transfer-type bioelectrocatalysis. Phys Chem Chem Phys 9(15):1793–1801
Khaldi H et al (2017) Efficiency of wastewater treatment by a mixture of sludge and microalgae. J Fundam Appl Sci 9:1454–1472
Kiely PD et al (2011) Anode microbial communities produced by changing from microbial fuel cell to microbial electrolysis cell operation using two different wastewaters. Bioresour Technol 102:388–394
Kiran R, Patil SA (2019) Microbial electroactive biofilms. In: Introduction to biofilm engineering. ACS Publications, pp 159–186
Koch C et al (2019) Trophic networks improve the performance of microbial anodes treating wastewater. npj Biofilms Microbiomes 5:27
Kolev Slavov A (2017) General characteristics and treatment possibilities of dairy wastewater—a review. Food Technol Biotechnol 55:14–28
Kothari R et al (2013) Production of biodiesel from microalgae Chlamydomonas polypyrenoideum grown on dairy industry wastewater. Bioresour Technol 144:499–503
Kouzuma A et al (2018) Metagenomic insights into the ecology and physiology of microbes in bioelectrochemical systems. Bioresour Technol 255:302–307
Kracke F et al (2015) Microbial electron transport and energy conservation–the foundation for optimizing bioelectrochemical systems. Front Microbiol 6:575
Karadag D, Köroğlu OE, Ozkaya B, Cakmakci M (2015) A review on anaerobic biofilm reactors for the treatment of dairy industry wastewater. Process Biochem 50(2):262–271
Kumar R et al (2017) Microbial fuel cells: types and applications. In: Waste biomass management—a holistic approach. Springer, Cham, pp 367–384
Kumari P et al (2022) Polyhydroxybutyrate production by Chlorella sorokiniana SVMIICT8 under nutrient-deprived mixotrophy. Bioresour Technol 354:127135
Logan BE et al (2019) Electroactive microorganisms in bioelectrochemical systems. Nat Rev Microbiol 17:307–319
Lovley DR (2012) Electromicrobiology. Annu Rev Microbiol 66:391–409
Luo X, Zhu J, Gleisner R, Zhan H (2011) Effects of wet-pressing-induced fiber hornification on enzymatic saccharification of lignocelluloses. Cellulose 18:1055–1062
Martinez CM, Alvarez LH (2018) Application of redox mediators in bioelectrochemical systems. Biotechnol Adv 36:1412–1423
Meena RAA et al (2019) Trends and resource recovery in biological wastewater treatment system. Bioresour Technol Rep 7:100235
Minteer SD, Liaw BY, Cooney MJ (2007) Enzyme-based biofuel cells. Curr Opin Biotechnol 18(3):228–234
Mohan SV, Saravanan R, Raghavulu SV, Mohanakrishna G, Sarma P (2008) Bioelectricity production from wastewater treatment in dual chambered microbial fuel cell (MFC) using selectively enriched mixed microflora: effect of catholyte. Bioresour Technol 99(3):596–603
Mohan SV, Nikhil GN, Chiranjeevi P, Reddy CN, Rohit M, Kumar AN, Sarkar O (2016) Waste biorefinery models towards sustainable circular bioeconomy: critical review and future perspectives. Bioresour Technol 215:2–12
Mohan SV et al (2010) Bio-catalyzed electrochemical treatment of real field dairy wastewater with simultaneous power generation. Biochem Eng J 51:32–39
Mohan SV et al (2019) Microbial electrochemical technology: emerging and sustainable platform. In: Microbial electrochemical technology. Elsevier, London, pp 3–18
Mohanakrishna G et al (2015) Bioelectrochemical systems (BES) for microbial electroremediation: an advanced wastewater treatment technology. In: Applied environmental biotechnology: present scenario and future trends. Springer, New Delhi, pp 145–167

Nancharaiah Y et al (2016) Biological and bioelectrochemical recovery of critical and scarce metals. Trends Biotechnol 34:137–155

Nancharaiah Y et al (2019) Removal and recovery of metals and nutrients from wastewater using bioelectrochemical systems. In: Microbial electrochemical technology. Elsevier, London, pp 693–720

Nuramkhaan M et al (2019) Isolation of microalgal strain from algal-bacterial aerobic granular sludge and examination on its contribution to granulation process during wastewater treatment in respect of nutrients removal, auto-aggregation capability and EPS excretion. Bioresour Technol Rep 8:100330

Okeke ES et al (2022) Microalgae biorefinery: an integrated route for the sustainable production of high-value-added products. Energy Convers Manage X 16:100323

Pandey P et al (2016) Recent advances in the use of different substrates in microbial fuel cells toward wastewater treatment and simultaneous energy recovery. Appl Energy 168:706–723

Pant D et al (2010) A review of the substrates used in microbial fuel cells (MFCs) for sustainable energy production. Bioresour Technol 101:1533–1543

Patil SA et al (2015) A logical data representation framework for electricity-driven bioproduction processes. Biotechnol Adv 33:736–744

Pocaznoi D et al (2012) Stainless steel is a promising electrode material for anodes of microbial fuel cells. Energy Environ Sci 5:9645–9652

Qin L et al (2014) Cultivation of Chlorella vulgaris in dairy wastewater pretreated by UV irradiation and sodium hypochlorite. Appl Biochem Biotechnol 172:1121–1130

Rabaey K, Rozendal RA (2010) Microbial electrosynthesis—revisiting the electrical route for microbial production. Nat Rev Microbiol 8:706–716

Rabaey K et al (2011) Metabolic and practical considerations on microbial electrosynthesis. Curr Opin Biotechnol 22:371–377

Ranadheer P, Kona R, Sreeharsha RV, Mohan SV (2019) Non-lethal nitrate supplementation enhances photosystem II efficiency in mixotrophic microalgae towards the synthesis of proteins and lipids. Bioresour Technol 283:373–377

Reguera G (2018) Microbial nanowires and electroactive biofilms. FEMS Microbiol Ecol 94(7):fiy086

Resch G et al (2008) Potentials and prospects for renewable energies at global scale. Energy Policy 36:4048–4056

Rousseau R et al (2013) Microbial bioanodes with high salinity tolerance for microbial fuel cells and microbial electrolysis cells. Electrochem Commun 33:1–4

Rozendal RA, Hamelers HV, Euverink GJ, Metz SJ, Buisman CJ (2006) Principle and perspectives of hydrogen production through biocatalyzed electrolysis. Int J Hydrog Energy 31(12):1632–1640

Rozendal RA, Leone E, Keller J, Rabaey K (2009) Efficient hydrogen peroxide generation from organic matter in a bioelectrochemical system. Electrochem Commun 11(9):1752–1755

Sabourin-Provost G, Hallenbeck PC (2009) High yield conversion of a crude glycerol fraction from biodiesel production to hydrogen by photofermentation. Bioresour Technol 100(14):3513–3517

Sadhukhan J et al (2014) Biorefineries and chemical processes. Design, integration and sustainability analysis. Wiley, Chichester

Sharma M et al (2014) A critical revisit of the key parameters used to describe microbial electrochemical systems. Electrochim Acta 140:191–208

Sreeharsha RV, Venkata Mohan S (2021) Symbiotic integration of bioprocesses to design a self-sustainable life supporting ecosystem in a circular economy framework. Bioresour Technol 326:124712

Sreeharsha RV, Tejaswini GS, Mohan SV (2022) Dual-stage biorefinery to convert spentwash hydrolysate into oleochemicals using Trichosporon cutaneum and Yarrowia lipolytica. Bioresour Technol 354:127146

Sreeharsha RV, Dubey N, Mohan SV (2023) Orienting biodiesel production towards sustainability and circularity by tailoring the feedstocks and processes. J Clean Prod:137526

Sivaramakrishnan R, Incharoensakdi A (2019) Enhancement of lipid extraction for efficient methyl ester production from Chlamydomonas sp. J Appl Phycol 31:2365–2377

Sleutels TH et al (2012) Bioelectrochemical systems: an outlook for practical applications. ChemSusChem 5:1012–1019

Sravan JS et al (2020) Cathodic selenium recovery in bioelectrochemical system: regulatory influence on anodic electrogenic activity. J Hazard Mater 399:122843

Stams AJ et al (2006) Exocellular electron transfer in anaerobic microbial communities. Environ Microbiol 8:371–382

Strik DP, Timmers RA, Helder M, Steinbusch KJ, Hamelers HV, Buisman CJ (2011) Microbial solar cells: applying photosynthetic and electrochemically active organisms. Trends Biotechnol 29(1):41–49

Timmers RA, Strik DP, Arampatzoglou C, Buisman CJ, Hamelers HV (2012) Rhizosphere anode model explains high oxygen levels during operation of a Glyceria maxima PMFC. Bioresour Technol 108:60–67

Ummalyma SB, Sukumaran RK (2014) Cultivation of microalgae in dairy effluent for oil production and removal of organic pollution load. Bioresour Technol 165:295–301

Vanthoor-Koopmans M et al (2013) Biorefinery of microalgae for food and fuel. Bioresour Technol 135:142–149

Velvizhi G, Mohan SV (2011) Biocatalyst behavior under self-induced electrogenic microenvironment in comparison with anaerobic treatment: evaluation with pharmaceutical wastewater for multi-pollutant removal. Bioresour Technol 102(23):10784–10793

Velvizhi G, Mohan SV (2012) Electrogenic activity and electron losses under increasing organic load of recalcitrant pharmaceutical wastewater. Int J Hydrog Energy 37:5969–5978

Wang X, Wen X, Yan H, Ding K, Zhao F, Hu M (2011) Bacterial community dynamics in a functionally stable pilot-scale wastewater treatment plant. Bioresour Technol 102(3):2352–2357

Wang P, Yu Z, Qi R, Zhang H (2016) Detailed comparison of bacterial communities during seasonal sludge bulking in a municipal wastewater treatment plant. Water Res 105:157–166

Wetser K, Sudirjo E, Buisman CJ, Strik DP (2015) Electricity generation by a plant microbial fuel cell with an integrated oxygen reducing biocathode. Appl Energy 137:151–157

Wilberforce T et al (2021) Value added products from wastewater using bioelectrochemical systems: current trends and perspectives. J Water Process Eng 39:101737

Yakushi T, Matsushita K (2010) Alcohol dehydrogenase of acetic acid bacteria: structure, mode of action, and applications in biotechnology. Appl Microbiol Biotechnol 86:1257–1265

Yang E et al (2019) Critical review of bioelectrochemical systems integrated with membrane-based technologies for desalination, energy self-sufficiency, and high-efficiency water and wastewater treatment. Desalination 452:40–67

Ye Y et al (2019) Microbial fuel cell for nutrient recovery and electricity generation from municipal wastewater under different ammonium concentrations. Bioresour Technol 292:121992

Chapter 6
Microalgae Photosynthesis for Nutrient Recovery and Value Addition

Abstract The current chapter discusses the potential applications of microalgae in various fields, including human health, nutrition, and biofuel production. Microalgae are single-cell photosynthetic eukaryotes that can grow in both fresh and marine environments, and they are capable of producing high-value chemicals and fuels. The carbon fixation process results in the biosynthesis of photosynthates, which can be channeled into different primary and secondary metabolites depending on the growth conditions and nutrient load. Microalgae are known to accumulate intracellular lipids, including exotic lipid groups such as eicosapentaenoic acid and docosahexaenoic acid which have multiple health benefits. They can also produce other high-value products such as natural dyes, polysaccharides, pigments, antioxidants, bioactive compounds, and proteins. The use of microalgae as a renewable feedstock has the potential to provide multiple benefits, including sustainable energy production, wastewater treatment, and the production of high-value chemicals. Chlorophyll and its derivatives have therapeutic properties and exhibit chemopreventive effects, while carotenoids play a vital role in preventing oxidative damage to cells and tissues. Microalgae are abundant sources of carotenoids including β-carotene, astaxanthin, lutein, fucoxanthin, zeaxanthin, antheraxanthin, violaxanthin, neoxanthin, loroxanthin, diadinoxanthin, diatoxanthin, and siphonein. Astaxanthin is the most industrially exploited ketocarotenoid with a high market value and exhibits higher antioxidant activity than other antioxidants. The chapter also highlights various health benefits of carotenoids, including retinal and visual dysfunctions, immune modulation, anticancer properties, and inhibiting cardiovascular diseases.

Keywords Microalgae · Nutraceuticals · Carotenoids · Biofuels · Therapeutic compounds · Lipids

R. V. Sreeharsha, S. Venkata Mohan, *Microbial Photosynthesis*,
https://doi.org/10.1007/978-981-97-1253-3_6

6.1 Introduction

Microalgae are single-cell photosynthetic eukaryotes that can grow in both fresh and marine environments, with over 300,000 species known to exist (Scott et al. 2010). The three most abundant classes of microalgae are the diatoms (Bacillariophyceae), the green algae (Chlorophyceae), and the golden algae (Chrysophyceae) (Alaswad et al. 2015). Microalgae are important green feedstocks that rely on photosynthesis to bring value addition to renewable waste resources. Compared to terrestrial plants that are the photosynthetic counterparts, microbes have higher productivities and lower doubling times, and they are amenable to metabolic remodeling for accumulation of large amounts of end products. We will discuss the concepts of metabolic engineering for specific end products in Chap. 8. Production of industrially important compounds using photosynthetic microbes offers an advantage that they do not demand high-quality agricultural land to grow. Readers appreciate the advances in bioreactors and other large-scale cultivation systems to minimize the land usage for these microbes in Chap. 6. The present chapter focuses on the wide range of products that are produced form microalgae and their potential applications in various domains including human health, nutrition, and biofuel production.

The carbon fixation process results in biosynthesizing the photosynthates which is the key component in metabolic flux channeling and biomass accumulation. Depending on the growth conditions and nutrient load, the photosynthates are channelized into different primary and secondary metabolites which have specific applications. For instance, numerous studies have identified the ability of microalgae to accumulate higher lipids under nutrient-stress conditions. The sucrose which is the primary end product of photosynthesis will be converted into fatty acids and stored in the microalgae cells for future needs. Similarly, light intensity plays a major role in the accumulation of more pigments that result in more photosynthetic efficiency. There are specific microalgae that can produce specific pigments including astaxanthin, zeaxanthin, and lutein that have commercial value. Irrespective of the metabolite that is produced, photosynthesis is the primary metabolic activity that plays a crucial role in metabolite accumulation and value addition to the microalgae bioprocess.

Using microalgae for the production of high-value chemicals and fuels has multiple advantages, as the process can be coupled with atmospheric CO_2 mitigation and wastewater treatment. Microalgae bioprocess mitigates the CO_2 from the atmosphere through the natural photosynthesis process and produces non-toxic and biodegradable intracellular storage compounds. Similarly, they can also use pollutants present in wastewater as nutrients, making them an eco-friendly technological medium for pollutant removal. Several studies have demonstrated the potential of microalgae in removing organic and inorganic pollutants including heavy metals, metalloids, and organic toxic compounds from wastewater (Christenson and Sims 2011; Gattullo et al. 2012). Furthermore, microalgae have the capability to synthesize several valuable substances like polyunsaturated fatty acids, natural colors,

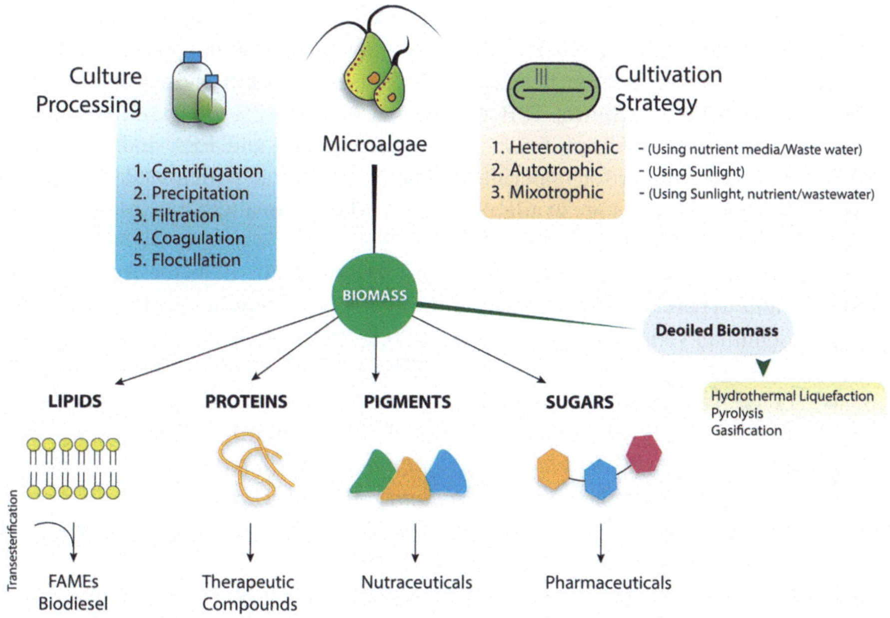

Fig. 6.1 Holistic view of microalgae cultivation and diverse groups of compounds produced from microalgae

polysaccharides, pigments, antioxidants, bioactive compounds, and proteins (Brennan and Owende 2010; Kiran and Venkata Mohan 2021; Mata et al. 2010). Hence, the utilization of microalgae as a sustainable resource has promise in yielding several advantages, including the generation of sustainable energy, treatment of wastewater, and synthesis of valuable compounds (Fig. 6.1).

6.2 Microalgae Biomolecules as Value-Added Compounds

6.2.1 *Lipids*

Microalgae have been seen to exhibit intracellular lipid accumulation, with lipid content varying between 30% and 70% of dry weight, contingent upon the specific species and development circumstances. Various factors, including nutritional content, temperature, light flow, and CO_2 concentration, exert significant influence on the process of lipid formation in microalgae. Certain microalgal species, such as *Phaeodactylum tricornutum*, *Nannochloropsis* sp., *Thraustochytrium*, and *Schizochytrium limacinum*, produce exotic lipid groups including eicosapentaenoic acid (EPA) and docosahexaenoic acid (DHA), which have multiple health benefits (Burja et al. 2006). EPA is a ω3 fatty acid that can replace membrane fluidity and prevent oxidative damage to neurons. EPA+DHA-rich algal oil capsules are sold as food supplements in the USA. Microalgae are a prominent natural source of DHA, which is a dietary supplement because of its nervous system and cardiovascular health benefits. Thraustochytrid 12B produces around 43.2% of DHA from 58% lipids biomass, while *Schizochytrium* sp. produces DHA and EPA and is widely used as food nourishment for maternity women and cardiac patients (Posser et al. 2018). DHA-rich algal oils are available in the market as an alternative to the use of fish oil (Long et al. 2018).

PUFA reduces the occurrence of obesity, rheumatism, diabetes, myocardial infarction, and hypercholesterolemia, and improves brain function, central nervous system, and human vision (Levasseur et al. 2020; Sathasivam et al. 2019). The lipids produced by microalgae contain a ratio of PUFA to saturated fatty acids at around 1.65 to 3.71, considered a balanced diet due to the presence of γ-linoleic acid (18:3), arachidonic acid (20:4), eicosapentaenoic acid (20:5), and docosahexaenoic acid (22:6) (Kumar et al. 2019). *Isochrysis galbana*, *Phaeodactylum tricornutum*, *M. subterraneus*, *Crypthecodinium cohnii*, *P. cruentum*, *C. calcitrans*, *Nannochloropsis* sp., *and P. salina* are PUFA producing microalgae species (Chiranjeevi and Mohan 2016; Mourelle et al. 2017).

Microalgal lipids are good sources of triacylglycerols which will produce fatty acid methyl esters with biodiesel properties upon transesterification. Some microalgae strains, including *Chlorella*, *Spirulina*, *Dunaliella*, and *Chlorococcus*, show considerable lipid accumulation when grown using wastewaters, suggesting the potential of microalgae for nutrient recovery of wastewaters (Chandra et al. 2015; Devi et al. 2012; Sabeela Beevi and Sukumaran 2015). The monounsaturated fatty acids (MUFA) and PUFA produced in microalgae have nutritional importance and serve as diet supplements for humans (Katiyar and Arora 2020; Rohit and Venkata Mohan 2018).

Microalgae have the ability to produce biomass and bioenergy in addition to treating wastewater in bio-photovoltaic cell systems, in addition to being valuable as food sources. The symbiotic relationship between oxygenic photosynthesis, specifically including microalgae as the cathode, and anoxygenic photosynthesis, involving photosynthetic/anaerobic bacteria as the anode, has the potential to yield bioenergy and biomass while also serving as a means of wastewater treatment. The use of microalgae as biocathodes eliminates the need for an external terminal electron acceptor, which is typically necessary to aid the oxygen reduction processes. The presence of microalgae engaging in oxygenic photosynthesis at the cathode contributes to the sustenance of elevated amounts of dissolved oxygen, so obviating the necessity for energy-demanding aeration practices.

Around 15% to 80% of the weight of microalgae is mainly composed of lipids. The lipid concentration can be influenced by factors such as the C/N (carbon to nitrogen) ratio in the medium, the induction of stress factors, and the culture conditions. Under situations of nitrogen restriction, microalgal cells exhibit an increased accumulation of lipids compared to nitrogen-rich environments. Numerous stress-inducing techniques have been investigated to increase lipid productivity, including high temperatures, nitrogen limitations, varied nutritional compositions, salt concentrations, and pH (Kwak et al. 2016). Appropriate stress conditions or other engineering techniques can boost microalgae's lipid output. In comparison to other oil-based crops, this particular crop has potential as a viable option for biodiesel generation, while also offering the additional benefit of functioning as a food supplement (Yeh and Chang 2012).

6.2.2 Proteins

Microalgae have a significant number of proteins, constituting around 42% to 70% of their dry cell weight. Nitrogen serves as the principal constituent of amino acids, chlorophyll, nucleic acids, and energy molecules such as ATP (Griffiths and Harrison 2009; Ranadheer et al. 2019). Several microalgae species such as *Chlorella*, *Arthrospira*, *Scenedesmus*, *Micractinium*, *Dunaliella*, *Chlamydomonas*, and *Euglena* are known for their protein content, and their proteins could serve as dietary supplements. Compared to animal-derived proteins, microalgae proteins require much less land, making them an environmentally sustainable choice. Microalgal biomass contains a substantial proportion of amino acids that are crucial for maintaining human health (Dawczynski et al. 2018). *Chlorella*, *Arthrospira* (Spirulina), *Scenedesmus*, *Micractinium*, *Dunaliella*, *Chlamydomonas*, *Euglena*, etc. are known for their rich protein content. *Spirulina* naturally called as single-cell protein (SCP) is recognized as a notable reservoir of phycobiliproteins, which are regarded as a valuable source of vitamin B1 and γ-linolenic acid (GLA). Microalgae proteins that are expressed either from chloroplast or nuclear genome could be a potential dietary supplement equivalent to high-quality protein foods such as β-lactoglobulin, tofu, mushrooms, soybean, and egg white (Caporgno and Mathys 2018; Hempel et al. 2011).

Microalgae proteins and protein derivatives have medicinal properties specific to humans, making them ideal candidates for drug development. For instance, cyanovirin produced by Nostoc sp. has demonstrated potential in curing human influenza A and HIV (Bhattacharjee 2016). Glycoproteins isolated from *Alexandrium minutum* stimulate mitophagy in alveolus cancer cell lines without harming healthy live cells. Microalgal proteins are involved in the synthesis of pancreozymins, which decrease appetite and lower LDL cholesterol (Galasso et al. 2019). Cosmetic and beauty products are made with protein extracts from microalgae, such as *Dermochlorella* DG, which stimulates collagen production and reduces vascular lesions (Levasseur et al. 2020). Microalgae also produce bioactive compounds called mycosporine-like amino acids (MAAs) that act as antioxidants and photostabilizers, shielding cells from UV-induced damage (Kiran and Venkata Mohan 2021). Peptides isolated from Chlorella vulgaris prevent diseases like atherosclerosis, cancer, and cardiovascular disorders, as well as protect DNA from oxidative damage, tumor metastasis, and immunosuppression (Ejike et al. 2017). Procollagen, a marker for hepatic fibrinolysis and liver disorders, was reduced by Pro-Gly-Trp-Asn-Gln--Trp-Phe-Leu and Val-Glu-Val-Leu-Pro-Pro-Ala-Glu-Leu peptides from the microalgae *N. incerta* (Kim and Kang 2011). Other peptides from *N. incerta* reduce Procollagen, a marker for hepatic fibrinolysis and liver disorders. Glycoproteins from *Chlorella vulgaris* protect mice against tumor spread and chemotherapy-induced immunosuppression (Andrade et al. 2018). Inhibiting the mitotic cell cycle at metaphase, Cryptophycin 1, a bioactive metabolite isolated from Nostoc species, demonstrated anti-proliferative and anti-mitotic effects via binding to the microtubule ends. When isolated from *Microcystis aeruginosa* and other microalgae, aeruginosins and depsipeptides (micropeptin, microcystilide, cyanopeptolin, oscillapeptin, and nostocyclin) exhibit inhibitory effects against plasmin, thrombin, trypsin, and chymotrypsin (Sathasivam et al. 2019).

Recombinant proteins have also been produced from *Chlamydomonas reinhardtii* and *Chlamydomonas elipsoidea*, such as vascular endothelial growth factor (VEGF), erythropoietin, IgG αPA83, hGAD65, immunotoxin αCD22-PE40, plasmodium surface proteins (AMS1, MSP1, VP28, Pfs25, Pfs28, CtxB-Pfs25, and Pfs 48/45), immunotoxins against B-cell lymphoma, and antibodies against anthrax, Herpes simplex virus, and human papillomavirus. These proteins have therapeutic applications in wound healing, anti-malarial vaccines, and anemia treatment, among others (Brasil et al. 2017; Hempel and Maier 2016).

6.2.3 Carbohydrates

Through photosynthesis, microalgae accumulate carbohydrates which constitute up to 55% of cell dry weight. The carbohydrates can be of storage or structural in the form of starch or cellulose, respectively (Chen et al. 2013). The carbohydrates derived from microalgae hvae inherent qualities that make them suitable for utilization in various conversion technologies. These carbohydrates are naturally occurring, biodegradable, non-toxic, and biocompatible. B-(1,3) glucan is a highly versatile immunostimulator derived from several sources. Chlorella has been found to mitigate the effects of hydroxyl radicals on cholesterol levels, as well as provide relief for constipation and hypercholesterolemia (Chen and Wang 2022). Microalgal-based polysaccharides are gaining importance in the cosmetic industry as hygroscopic agents and antioxidants for topical applications (Dragone et al. 2011). Phycobiliproteins are an integral component of the microalgae membrane and light-harvesting complex. The extracellular polysaccharides synthesized from Rhodella reticulate showed antioxidant activity. Animals fed with polysaccharide-rich algae lowered cholesterol levels in serum with the subsequent reduction in insulin and glucose. Polysaccharide extract (10 mg kg^{-1}) from the *Porphyridium cruentum* showed anti-proliferative activity and was able to mitigate myeloid tumor proliferation in hamsters in vivo (Gardeva et al. 2009).

Certain seaweed species of Rhodophyta, Phaeophyta, Chlorophyta divisions produce sulfate esters of polysaccharides which have anticancer, antihepatotoxicity, and fungicidal properties. The sulfated polysaccharides including ulvan, carrageenan, and fucoidan produced by microalgae are known to inhibit viral infection by blocking viral adsorption on host cells (de Jesus Raposo et al. 2014; Pereira et al. 2018). Sulfated polysaccharides isolated from *R. reticulate* show strong antioxidant property against superoxide anions. Moreover, the efficacy of sulfated polysaccharides derived from *Scenedesmus rubescens*, *Scenedesmus acuminatus*, and *Phaeodactylum tricornutum* in the inhibition of alpha-glucosidase, glutaminyl-peptide cyclotransferase, and telomerase activities has been demonstrated (Olasehinde et al. 2017). It is noteworthy that the sulfated polysaccharides derived from Porphyridium sp. have the ability to function as a coating agent on sanitary products, so serving as a preventive measure against the transmission of infectious illnesses of epidemic nature (Gaikwad 2022).

6.2.4 Chlorophyll

Chlorophyll, an essential pigment for photosynthesis, is present in all plants, algae, and cyanobacteria and exhibits solubility in lipids. Due to its green pigmentation and ability to synthesize natural products, chlorophyll has gained importance as a dye agent in the food and pharmaceutical industries. *Chlorella* Sp. is a type of green microalgae that is known as "Emerald food" due to its high chlorophyll content (Khanra et al. 2018; Odjadjare et al. 2017). Chlorophyll and its derivatives have antioxidant and therapeutic properties, making them useful in fighting oxidation. Chlorophyll a, b, or their combination has been found to demonstrate chemopreventive effects. These effects include the elevation of glutathione S-transferase levels, inhibition of cytochrome P450 enzymes, suppression of cellular differentiation, promotion of mitotic arrest, and induction of necrobiosis (Mishra et al. 2011). Research shows that chlorophyll in combination with lutein can arrest hepatoma cells in the G0/G1 phase and induce necrosis or apoptosis. Chlorophyll from *Sauropus androgynous* has potent antioxidant activity in vivo when administered prooxidant sodium nitrate orally to Wistar rats for 14 days (Suparmi et al. 2016). Chlorophyllin, a derivative of chlorophyll, has high effectiveness as a chemopreventive agent against HCT1 16 human colon cancer cells (Koller et al. 2014). The compound finds extensive use in various health and hygiene products, including but not limited to antiperspirants, air fresheners, lozenges, and formulations designed to combat unpleasant odors.

6.2.5 Carotenoids

Carotenoids are a class of hydrophobic accessory pigments characterized by a C40 backbone structure composed of isoprene units. They have a notable impact on the suppression of oxygen free radicals, functioning as antioxidants, and mitigating oxidative damage to cells, tissues, and membranes. In 2017, the projected global market value of carotenoids was 1.5 billion. The antioxidant action can be ascribed to the existence of conjugated double bonds or the presence of epoxy, acetyl, allene, or acetylene groups. There exists a diverse array of over 400 carotenoids that have been found across numerous organisms. Among these, the most often seen carotenoids in commercial applications are β-carotene, astaxanthin, lutein, fucoxanthin, zeaxanthin, antheraxanthin, violaxanthin, neoxanthin, loroxanthin, diadinoxanthin, diatoxanthin, and siphonein. Carotenoids find extensive use in the manufacturing of nutritional supplements, nutritious food products, aquaculture practices, and animal feed formulations (Galasso et al. 2019). Several species of *Dunaliella* are abundant sources of carotenoids and have exhibited antihyperlipidemic activities (El-Baky et al. 2009). Lycopene synthesized from *Chlorella* species has been shown to reduce blood cholesterol, inhibit the growth of prostate cancer in mice, reduce lipoprotein levels, and enhance rheumatoid arthritis. Similarly, *C. ellipsoidea* and *C. vulgaris* rich in carotenoids have been found to inhibit the proliferation of colon cancer cells (de Morais et al. 2015). The first nutritional product produced commercially from the genera *Dunaliella* was β-carotene, which is a precursor of Vitamin A and contributes several benefits to human welfare, such as improving retinal and visual dysfunctions, immune modulation, anticancer, inhibiting cardiovascular diseases, and delaying aging (Barkia et al. 2019). Oral intake of 35 to 185 mg of β-carotene per day has been reported to suppress UV-induced erythema in living beings. β-carotene plasma levels have been inversely related to atherosclerotic lesions, influencing plasma triglycerides, cholesterol, high-density lipoprotein, and inhibiting low-density lipoprotein oxidation. The cognitive impairment associated with Alzheimer's disease in rats has been seen to be greatly inhibited by the administration of lutein and β-carotene derived from *Chlorella* sp. (de Morais et al. 2015). Several dominant microalgal species rich in β-carotene include *Chlorella sorokiniana*, Chlamydocapsa sp., *Dunaliella salina*, *Chlorococcum* sp., *Dunaliella tertiolecta*, *B. braunii*, *Chlamydomonas nivalis*, and *Paeonia obovate* (Gong and Bassi 2016). Astaxanthin, a xanthophyll pigment of red coloration, holds the distinction of being the ketocarotenoid that is most extensively utilized in industrial applications. This compound possesses a significant market value, estimated to be around $200 million. Astaxanthin demonstrates significantly elevated antioxidant activity when exposed to reactive oxygen species, surpassing the levels observed in other compounds. Specifically, its antioxidant activity is approximately 100 times greater than that of α-tocopherol, 6000 times greater than vitamin C, 800 times higher than coenzyme Q10, 550 times greater than vitamin E, 200 times higher than polyphenols, 150 times greater than anthocyanins, and 75 times higher than α-lipoic acid (Talukdar et al. 2020).

The addition of astaxanthin as a potential adjunctive provides resilient protection against swelling, cancer, syndrome X, and neurodegenerative and ocular disorders (Galasso et al. 2019). The inhibitory effects of astaxanthin on oxidative stress were observed, accompanied by the activation of quinone oxidoreductase (NQO-1), glutathione-S-transferase-α1 (GST-α1), and heme oxygenase (HO-1) expression. These effects were mediated through the antioxidant response element (Nrf2-ARE) and the nuclear factor erythroid-related factor 2 (Nrf2) signaling pathway. Astaxanthin has been found to have protective effects on pancreatic beta cells through its ability to mitigate oxidative stress and reduce sugar toxicity. Additionally, it has been observed to enhance adiponectin and HDL levels, as well as promote improved blood flow and circulation (Li et al. 2020).

Astaxanthin has been found to have an antagonist effect on Helicobacter pylori, the bacteria that cause ethanol-induced gastritis in rodents. Additionally, the intake of astaxanthin has shown a significant impact against ethanol-induced gastric ulcers. Astaxanthin has the potential to mitigate the onset of acute respiratory distress syndrome (ARDS) and sepsis generated by lipopolysaccharide. This is achieved through the inhibition of the MPAK/NF-KB signaling pathway and the subsequent reduction of pro-inflammatory cytokine levels. Similarly, as a dietary supplement, astaxanthin stimulates mitogen-induced lymphocyte proliferation, increases natural killer cell cytotoxicity, the number of total T and B cells in the peripheral blood, and the delayed-type hypersensitivity response. Additionally, the study demonstrated protective effects against lung damage, suppression of alveolar wall swelling, reduction in myeloperoxidase activity, and mitigation of the drop in the quantity of pulmonary alveoli within lung tissues (Capelli et al. 2019). Astaxanthin provides protection to brain cells against apoptosis triggered by Aβ25–35 through the inhibition of radical oxidative damage, Bax production, nuclear transcription factor-kB (NFkB), and dephosphorylation of P-38 MAP kinase (mitogen-activated protein). The potential efficacy of natural astaxanthin in combating COVID-19 is supported by several preclinical and human trials, which have demonstrated its anti-inflammatory and antioxidant properties. Furthermore, its exceptional safety profile further enhances its potential as a viable therapeutic candidate for the treatment of COVID-19.

Lutein is a yellow carotenoid that mainly protects photoreceptors by filtering blue light that damages the lens and retina of the eyes. Microalgae produce 70–150 tons/ha of lutein annually, which is 11.5–25 times higher than marigold flowers (6 tons) with a market value of $3.14 million (Lin et al. 2015). Lutein accumulates in the macula, a part of the human retina, and supplementary intake prevents macular degeneration that occurs among older people (Chiu et al. 2016). In cell culture tests, it has been observed that lutein demonstrates superior antioxidant activity compared to β-carotene. This is attributed to its ability to mitigate oxidative damage and protect lipid bilayers from harm. Epidemiological investigations have provided evidence suggesting that lutein exhibits efficacy in postponing the initiation of chronic ailments (Rasmussen and Johnson 2013). Daily intake of 6 mg of lutein in combination with zeaxanthin prevents glaucoma, cataracts, low vision, atherosclerosis, and age-related macular degeneration. It also protects endothelial cells in vitro and possesses anticancer activity (Liu et al. 2017). The resistance of human cells to a 30 mM concentration of H_2O_2 was seen when treated with a hydroalcoholic extract containing antheraxanthin, neoxanthin, carotenes, lutein, and violaxanthin derived from *Tetraselmis suecica*. (Sansone et al. 2017). An in vivo study demonstrated that a mixture of lutein (3.7 mg), zeaxanthin (0.8 mg), and meso-zeaxanthin (7.3 mg) produced an enhancement of macular pigments in humans. The enriched extract (0.17 μM) showed dose-dependent growth inhibition of human breast adenocarcinoma cell lines (MCF-7) (Connolly et al. 2010).

Violaxanthin, a member of xanthophyll and orange carotenoid showed intense anti-proliferative activity on human mammary cancer cells (Koller et al. 2014). The hydroalcoholic extract containing antheraxanthin, neoxanthin, carotenes, lutein, and violaxanthin derived from Tetraselmis suecica exhibited a notable level of resistance in human cells upon exposure to a concentration of 30 mM of hydrogen peroxide (H_2O_2) (Sampathkumar and Gothandam 2019). Zeaxanthin and fucoxanthin are utilized in the pharmaceutical and cosmetic sectors, primarily recognized for their anticancer and anti-proliferative attributes. Moreover, fucoxanthin has been found to possess preventive properties against bone ailments, including osteoporosis, rheumatoid arthritis, and diabetes conditions characterized by insulin suppression and hyperglycemia (Gong et al. 2018).

6.2.6 Vitamins

Microalgae are a rich source of vitamins that are widely used as supplements to enhance cellular function, detoxify the body, and activate the immune system. Microalgae are known to generate several vitamins, including vitamin A (β-carotene), vitamin C, vitamin E, and a range of B vitamins such as thiamine (B1), riboflavin (B2), niacin (B3), pantothenic acid (B5), pyridoxine (B6), folic acid (B9), and cyanocobalamin (B12). Brown microalgae, *Eisenia arborea*, contains 0.34 mg g^{-1} of vitamin C in dry biomass, while Chlorella is a rich source of vitamin A, which plays a critical role in vision, reproduction, immune function, and cellular communication. Vitamin E synthesized by microalgae such as *D. tertiolecta*, *T. suecica*, *Nannochloropsis* sp., *C. calcitrans*, and *Porphyridium* sp. acts as a shield to protect the lipid bilayers from reactive oxygen species and avert atherosclerotic heart disease and neurodegenerative diseases (Andrade et al. 2018). Vitamin B3, which is essential for fat metabolism, blood glucose regulation, cholesterol biosynthesis, nucleic acid metabolism, and coronary heart diseases, is abundantly biosynthesized in microalgae. Vitamin B1 has anti-inflammatory activity, while vitamin B7 helps in retaining hair follicles, fingernails, and skin regeneration. Microalgae such as *Chlorella* and Spirulina are a rich source of folic acid (vitamin B9), which is necessary for protecting skin and mucous-secreting glands, cellular metabolism, and maintaining healthy teeth and bones. Marine algae-derived Vitamin B12 is employed as an encapsulation agent for antineoplastic medications and assumes a crucial function in histone methylation, DNA replication, and hematopoiesis. The presence of elevated levels of cobalamin and folate has been associated with a reduced risk of developing breast cancer (Delasoie et al. 2018). Vitamin C functions as an immunomodulatory agent in the prevention of the dissemination of infectious illnesses, such as TB. Moreover, microalgae possess vitamin D2 and D3, which, in conjunction with provitamin D3, serve as chemopreventive agents, and exhibit anti-proliferative and immunomodulatory effects on tumor cells (Giammanco et al. 2015).

6.3 Therapeutic Applications of Microalgae

A wide range of compounds derived from microalgal biomass and their diverse applications are listed in Table 6.1. Carotenoids, polyunsaturated fatty acids (PUFAs), phycobiliproteins, carbohydrates, and sulfurized polysaccharides derived from microalgae have been identified for their ability to mitigate chronic inflammation and provide protective effects whether ingested as dietary components or used as medicinal supplements. Particularly, polyunsaturated fatty acids (PUFAs) such as arachidonic, docosahexaenoic, and eicosapentaenoic acids have been implicated in the therapeutic management of rheumatoid arthritis and dermatitis (Blagojević et al. 2018). Enzymatic and non-enzymatic antioxidants function as antineoplastic drugs by suppressing the action of cancer cell proteases, promoting cellular proliferation, inducing apoptosis, and modulating the immune response (Kim and Kang 2011). Algal extracts possess potential for the development of novel pharmaceuticals targeting bacterial illnesses, including those caused by foodborne and human pathogenic bacteria. The antibacterial action is ascribed to the presence of monounsaturated and polyunsaturated fatty acids. The oxylipins produced from microalgae have significant antibacterial efficacy against H. influenzae (with an IC50 value of 1.9 μg/mL) and methicillin-resistant Staphylococcus aureus (with an IC50 value of 7.8 μg/mL). The growth of several bacteria, such as E. coli, Pseudomonas aeruginosa, *Aeromonas hydrophila*, *Staphylococcus epidermidis*, *Photobacterium phosphoreum*, *Planococcus citreus*, and *Micrococcus luteus*, is hindered (Mostafa 2012). The major focus in current research involves the screening of microalgae for antiviral chemicals, with Cyanobacteria emerging as a possible source of industrial-scale antiviral medicines (Vijayakumar and Menakha 2015). Various viruses are known to infect humans and animals. These include human α herpesvirus 3 (HHV-3), herpes simplex virus (HSV1/HSV2), human β herpesvirus, murine leukemia virus (MuLV), Flu-A viruses, hepatitis B virus (HBV), viral hemorrhagic septicemia virus (VHSV), African swine fever virus (ASFV), vaccinia virus (VACV), vesicular stomatitis virus (VSV), encephalomyocarditis virus, flu-A and flu-B, respiratory syncytial virus type A (RSV-A) and B (RSV-B), measles, and mumps. Marine microalgae exhibit more potential in the development of innovative antifungal compounds as compared to their counterparts in freshwater environments.

6.4 Biofuels Applications of Microalgae

The identification and characterization of potential strains for biodiesel production are promising strategies nevertheless, the bioprocesses are suffering to reach theoretical lipid yields due to various reasons including growth inhibitors, sugar specificity, and feedback regulation of metabolic pathways (Sreeharsha et al. 2021, 2022). Furthermore, currently available biomass processing and biofuel production methods hinder the economic performance of algal-based biofuel biorefinery. It is

Table 6.1 Wide range of products derived from microalgae and their therapeutic applications, specific effect along with the microalgae species used as feedstock

Product	Organism produced	Effect	References
Anti-inflammatory			
Docosapentaenoic acid (DPA)	*Schizochytrium* sp.	Inhibits the production of cytokines (IL-1 and TNF-) in human peripheral blood mononuclear cells (PMBC)	Nauroth et al. (2010)
Astaxanthin	*Haematococcus pluvialis*	Possessed inhibitory effects on lipopolysaccharide-induced inflammation, and anti-inflammatory activity was more remarkable than prednisolone	Ohgami et al. (2003), Park et al. (2010)
Fucoidan	Brown algae	Phagocytic cells were inhibited from producing nitric oxide synthase and prostaglandins	Kang et al. (2011)
C. stigmatophora and P. tricornutum extracts	Red and brown seaweed, respectively	Showed positive responses against paw edema induced by carrageenan	Guzmán et al. (2003)
Fucoxanthin	Diatoms	Capable of eliciting metastasis in humans	Zhao et al. (2019)
Phycobiliproteins	Cyanobacteria	Inhibits histamine release	Zhao et al. (2019)
Spirulan	*A. platensis*	Identical to glycosaminoglycans (GAG) found in animals and involved in human physiological and pathological responses	Senni et al. (2011)
Exopolysaccharides	*Porphyridium cruentum*	Enhanced immunostimulating activity in vitro and inhibits membrane peroxidation	Sun et al. (2012)
Anti-cancer			
Cyanobacteria-derived natural products	Anabaena and Nostoc sp.	Inhibiting tumor cell growth, inducing programmed cell death	Burja et al. (2001)
Aqueous extract of Nostoc muscorum	*Nostoc muscorum*	Anticancer activity against Ehrlich cell (87.25%) and hepatocellular carcinoma (89.4%)	Shanab et al. (2012)
Dragonamide A-E and dragomabin	Lyngbya sp.	Best differential toxicity between the mammalian cells and parasites and in vitro activity against leishmaniasis	Singh et al. (2017)
Cryptophycin 52	Nostoc sp. GSV 224	Treating ovarian carcinoma and alveoli cancers in lab trials	Singh et al. (2017)
c-phycocyanin	Red algae	Exhibits antiproliferation inducing apoptosis	Wang et al. (2007)

(continued)

Table 6.1 (continued)

Product	Organism produced	Effect	References
Sulfated polysaccharide (p-KG03)	*G. impudicum*	Preventing tumor cell growth, rejuvenating the immune system, and enhancing cytokine production in macrophages	Bae et al. (2006)
β-(1,3)-glucan	*C. vulgaris*	Active immunostimulator, potential antitumor agent	Laroche and Michaud (2007)
Sulfated polysaccharides	*P. cruentum*	Intense antitumor activity in rodent peritoneum, impeding the progression of cancer cells, increasing spleen and thymus indices, and enhancing the immune system	Sun et al. (2012)
Exopolysaccharides	*Porphyridium* sp.	Retarding neoplasia, preventing proliferation of colon cancer in rats	Geresh et al. (2002)
Microalgae-derived peptides	Not specified	Potential treatment for gastric cancer	Sheih et al. (2010)
Antibacterial			
Parsiguine	*Fischerella ambigua*	MIC values of 40 mg/mL against S. epidermidis PTCC 1114 and 20 mg/mL against *C. krusei* ATCC 44507	Najdenski et al. (2013)
Antibacterial compounds	Microalgae species from different taxonomic groups including classes Chrysophytes and Diatomophyceae	Possess antibacterial activity against several human pathogenic bacteria including Escherichia coli, Pseudomonas aeruginosa, Enterobacter aerogenes, salmonella typhimurium, Klebsiella pneumonia, Vibrio cholera, and Proteus vulgaris	Falaise et al. (2016), Mudimu et al. (2014)
PUFA including eicosapentaenoic acid (EPA), hexadecatrienoic acid (HTA), and palmitoleic acid (PA).	*Chlorococcum* strain HS-101, *Dunaliella primolecta*, and diatom *Phaeodactylum tricornutum*	Exhibit significant bacteriostatic activity against methicillin-resistant Staphylococcus aureus (MRSA)	Desbois et al. (2009)
Short-chain fatty acids	*H. pluvialis*	Exhibit antibacterial activity towards E. coli	Rodríguez-Meizoso et al. (2010)
Long-chain fatty acids	*Scenedesmus obliquus*	Show antibacterial activity against Staphylococcus aureus	Rodríguez-Meizoso et al. (2010)

(continued)

Table 6.1 (continued)

Product	Organism produced	Effect	References
Polysaccharides	Microalgae	Can inhibit bacterial adhesion and aggregation properties. Act similar to E. coli polysaccharides	Rendueles and Ghigo (2012)
Ethanolic and methanolic extracts	*S. grantiana* and *O. sancta*	Show antibacterial activity against three human pathogens: E. coli, P. vulgaris, and P. mirabilis with the zone of inhibition of 9, 10 and 9 mm	Prakash et al. (2011)
Antiviral			
Polysaccharides	Marine microalgae	Broad antiviral spectrum against enveloped viruses	Vo et al. (2011)
Acyclovir®	*Dunaliella* sp.	Inactivates the viral function of HSV and HIV-1	Falaise et al. (2016)
Polysaccharides (PS)	*A. platensis*, *R. reticulate*, *Porphyridium* sp., *G. impudicum*, and *C. polykrikoides*	Notable microalgae with antiviral properties against a variety of viruses	Santoyo et al. (2011), Silva et al. (2018), Talyshinsky et al. (2002)
Sulfated polysaccharide p-KG03	*G. impudicum*	Good activity against the encephalomyocarditis RNA virus (EMCV) comparable to existing drugs	Kim et al. (2012)
Navicular and A1/A2	*N. directa* and *C. polykrikoides*	Strong antiviral activity against HIV1, HSV1, and type A flu viruses	Lee et al. (2006)
Endocellular extracts	*Porphyridium* sp., *D. tertiolecta*, *Tisochrysis lutea*, *I. galbana*, and *Chlorella*	Inhibited replication of viral hemorrhagic septicemia virus in vitro	Fabregas et al. (1999)
Antifungal			
Antifungal compounds	*Chaetoceros* sp.	Inhibit fungal growth	Alangaden (2011), Mudimu et al. (2014)
Pigments chlorophyll a, b and ά- and β-carotene	*C. humicola*, *Porphyridium* sp.	Antifungal properties	Bhagavathy et al. (2011)
Amphidinol	*Amphidinium klebsii*	Inhibit fungal growth by targeting fungi membranes, mainly ergosterols biosynthetic pathway	Ghannoum and Rice (1999)
Pigments	*H. karadagensis*	Antifungal activity against C. maritima, D. salina, and Lulworthia sp.	Gastineau et al. (2012)

(continued)

Table 6.1 (continued)

Product	Organism produced	Effect	References
Organic extracts	*C. lauderi*, *G. toxicus*, *C. vulgaris*	Inhibit Aspergillus fumigatus	Ghasemi et al. (2007)
ACE-inhibitory peptides	*Chlorella vulgaris*, *Nannochloropsis oculata*, *C. ellipsoidea*, and *Arthrospira platensis*	Lower blood pressure	Samarakoon et al. (2013)
		Block Angiotensin II production and prevent organ damage	
Nanofibrils	*A. platensis*	Used for treating spinal cord injuries	Raposo et al. (2010)
		Used as an extracellular matrix for culturing stem cells	
Spirulan	*A. platensis*	Demonstrates antithrombogenic properties. Platelet aggregation is inhibited by interfering with the blood coagulation-fibrinolytic system	Hayakawa et al. (2000)
Sulfated polysaccharides			
Carrageenans	Various seaweeds	Exhibits anticoagulant activity	Pereira et al. (2018)
		Inhibits platelet aggregation	

anticipated that advancements in the cultivation of microalgae and subsequent processing techniques, including harvesting, drying, and thermochemical processing, would enhance the economic viability of microalgae-derived biofuels to a certain degree (Li et al. 2008). A compendium of biodiesel production and the elements involved at various stages are represented in Fig. 6.2.

Petro-based fossil fuel resources are diminishing gradually, and their usage is posing a threat to the global climate. Microalgae-derived lipids are considered an alternative and renewable source to produce environmentally safe and clean biofuel, thereby reducing greenhouse gas emissions. Microalgae are classified as third-generation biofuels due to their notable benefits in terms of accelerated growth and enhanced productivity compared to first-generation feedstock, consisting of food crops, and second-generation feedstock, which mostly encompasses non-edible oil-producing tree species (Sreeharsha and Mohan 2020). Microalgae possess relatively high photosynthetic conversion efficiencies, diverse metabolic capabilities, superior growth rates, and the ability to store or secrete energy-rich hydrocarbons. In microalgae, cellular lipids serve both as energy storage molecules and to stabilize biological membranes. Fatty acids in the form of triacylglycerols are ideal for producing biodiesel through transesterification (Sreeharsha et al. 2023). Numerous microalgae have the capacity to synthesize significant quantities of triacylglycerols (TAG) as a kind of lipid storage in response to photo-oxidative stress or unfavorable environmental circumstances, earning them the designation of oleaginous algae. The oil content of different microalgae species exhibits variations relative to their dry weight, with certain species demonstrating oil concentrations that may reach up to 80% of their dry biomass. Microalgae has considerable potential as a viable and attractive alternative for biofuel production.

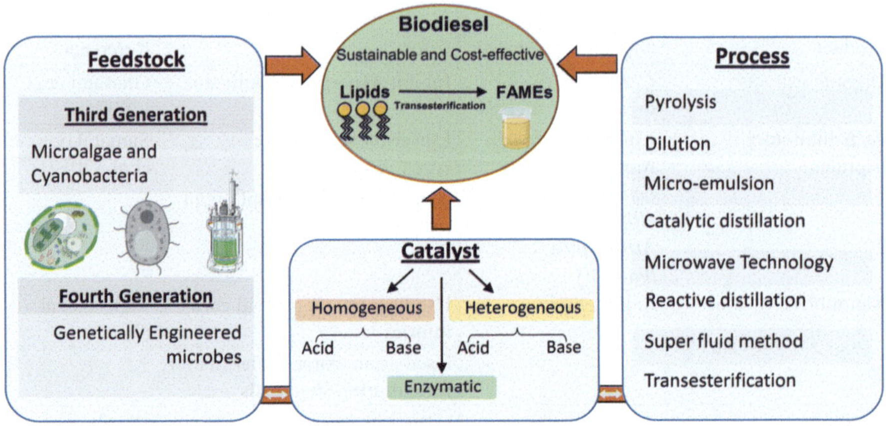

Fig. 6.2 A brief description of the process flow of biodiesel production and the elements involved at various stages

6.5 Conclusions

Microalgae represent a versatile group of single-cell photosynthetic organisms with immense potential across various domains. As green feedstocks, they are exceptional at carbon fixation and photosynthesis, through which they produce a diverse array of primary and secondary metabolites including lipids, proteins, carbohydrates, chlorophyll, carotenoids, and vitamins. The lipids can be converted into biodiesel, offering a sustainable and renewable alternative to fossil fuels. Additionally, microalgae-derived lipids contain essential polyunsaturated fatty acids like EPA and DHA, which have significant health benefits, further expanding their value. Proteins from microalgae can serve as dietary supplements and also have medicinal properties with applications in treating influenza, cancer, and other diseases. Carbohydrates produced by microalgae have applications in various conversion technologies and the cosmetic industry. Sulfate esters of polysaccharides, such as ulvan and fucoidan, exhibit anticancer, antihepatotoxicity, and antiviral properties. Chlorophyll and carotenoids derived from microalgae are used as natural colorants and antioxidants in the food and pharmaceutical industries. These pigments have therapeutic properties and are essential for human health. Microalgae are rich sources of vitamins, including vitamins A, C, E, and various B vitamins. These vitamins play crucial roles in maintaining health and are used as supplements to enhance cellular function, detoxify the body, and boost the immune system.

References

Alangaden GJ (2011) Nosocomial fungal infections: epidemiology, infection control, and prevention. Infect Dis Clin 25(1):201–225

Alaswad A et al (2015) Technologies and developments of third generation biofuel production. Renew Sust Energ Rev 51:1446–1460

Andrade L et al (2018) Chlorella and spirulina microalgae as sources of functional foods. Nutraceutical Food Suppl 6:45–58

Bae SY, Lee HK, Pyo S, Yim JH (2006) Activation of murine peritoneal macrophages by sulfated exopolysaccharide from marine microalga Gyrodinium impudicum (strain KG03): involvement of the NF-κB and JNK pathway

Barkia I et al (2019) Microalgae for high-value products towards human health and nutrition. Mar Drugs 17:304

Bhagavathy S, Sumathi P, Bell IJS (2011) Green algae Chlorococcum humicola-a new source of bioactive compounds with antimicrobial activity. Asian Pac J Trop Biomed 1(1):S1–S7

Bhattacharjee M (2016) Pharmaceutically valuable bioactive compounds of algae. Asian J Pharm Clin Res 9:43

Blagojević D et al (2018) Antioxidant activity and phenolic profile in filamentous cyanobacteria: the impact of nitrogen. J Appl Phycol 30:2337–2346

Brasil BSAF et al (2017) Microalgae and cyanobacteria as enzyme biofactories. Algal Res 25:76–89

Brennan L, Owende P (2010) Biofuels from microalgae—a review of technologies for production, processing, and extractions of biofuels and co-products. Renew Sust Energ Rev 14:557–577

Burja AM, Banaigs B, Abou-Mansour E, Burgess JG, Wright PC (2001) Marine cyanobacteria—a prolific source of natural products. Tetrahedron 57(46):9347–9377

Burja AM et al (2006) Isolation and characterization of polyunsaturated fatty acid producing Thraustochytrium species: screening of strains and optimization of omega-3 production. Appl Microbiol Biotechnol 72:1161–1169

Capelli B et al (2019) Astaxanthin sources: suitability for human health and nutrition. Funct Food Health Dis 9:430–445

Caporgno MP, Mathys A (2018) Trends in microalgae incorporation into innovative food products with potential health benefits. Front Nutr 5:58

Chandra R et al (2015) Lipid metabolism in response to individual short chain fatty acids during mixotrophic mode of microalgal cultivation: influence on biodiesel saturation and protein profile. Bioresour Technol 188:169–176

Chen H, Wang Q (2022) Microalgae-based green bio-manufacturing—how far from us. Front Microbiol 13:832097

Chen C-Y et al (2013) Microalgae-based carbohydrates for biofuel production. Biochem Eng J 78:1–10

Chiranjeevi P, Mohan SV (2016) Critical parametric influence on microalgae cultivation towards maximizing biomass growth with simultaneous lipid productivity. Renew Energy 98:64–71

Chiu P-H et al (2016) Cultivation of two thermotolerant microalgae under tropical conditions: influences of carbon sources and light duration on biomass and lutein productivity in four seasons. Bioresour Technol 212:190–198

Christenson L, Sims R (2011) Production and harvesting of microalgae for wastewater treatment, biofuels, and bioproducts. Biotechnol Adv 29:686–702

Connolly EE et al (2010) Augmentation of macular pigment following supplementation with all three macular carotenoids: an exploratory study. Curr Eye Res 35:335–351

Dawczynski C et al (2018) Docosahexaenoic acid in the treatment of rheumatoid arthritis: a double-blind, placebo-controlled, randomized cross-over study with microalgae vs. sunflower oil. Clin Nutr 37:494–504

de Jesus Raposo MF et al (2014) Bioactivity and applications of polysaccharides from marine microalgae. Springer, Cham

de Morais MG et al (2015) Biologically active metabolites synthesized by microalgae. BioMed Res Int 2015:835761
Delasoie J et al (2018) Slow-targeted release of a ruthenium anticancer agent from vitamin B 12 functionalized marine diatom microalgae. Dalton Trans 47:17221–17232
Desbois AP, Mearns-Spragg A, Smith VJ (2009) A fatty acid from the diatom Phaeodactylum tricornutum is antibacterial against diverse bacteria including multi-resistant Staphylococcus aureus (MRSA). Mar Biotechnol 11(1):45–52
Devi MP, Subhash GV, Mohan SV (2012) Heterotrophic cultivation of mixed microalgae for lipid accumulation and wastewater treatment during sequential growth and starvation phases: effect of nutrient supplementation. Renew Energy 43:276–283
Dragone G et al (2011) Nutrient limitation as a strategy for increasing starch accumulation in microalgae. Appl Energy 88:3331–3335
Ejike CE et al (2017) Prospects of microalgae proteins in producing peptide-based functional foods for promoting cardiovascular health. Trends Food Sci Technol 59:30–36
El-Baky H et al (2009) Production of phenolic compounds from Spirulina maxima microalgae and its protective effects. Afr J Biotechnol 3:133–139
Fabregas J, Garcıa D, Fernandez-Alonso M, Rocha AI, Gómez-Puertas P, Escribano JM et al (1999) In vitro inhibition of the replication of haemorrhagic septicaemia virus (VHSV) and African swine fever virus (ASFV) by extracts from marine microalgae. Antivir Res 44(1):67–73
Falaise C, François C, Travers MA, Morga B, Haure J, Tremblay R et al (2016) Antimicrobial compounds from eukaryotic microalgae against human pathogens and diseases in aquaculture. Mar Drugs 14(9):159
Gaikwad M (2022) Sulfated polysaccharide from marine red microalga porphyridium against SARS-CoV-2—a mini-review. Acta Sci Microb 5:128–136
Galasso C et al (2019) Microalgal derivatives as potential nutraceutical and food supplements for human health: a focus on cancer prevention and interception. Nutrients 11:1226
Gardeva E et al (2009) Cancer protective action of polysaccharide, derived from red microalga Porphyridium cruentum—a biological background. Biotechnol Biotechnol Equip 23:783–787
Gastineau R, Pouvreau JB, Hellio C, Morançais M, Fleurence J, Gaudin P et al (2012) Biological activities of purified marennine, the blue pigment responsible for the greening of oysters. J Agric Food Chem 60(14):3599–3605
Gattullo CE et al (2012) Removal of bisphenol A by the freshwater green alga Monoraphidium braunii and the role of natural organic matter. Sci Total Environ 416:501–506
Geresh S, Adin I, Yarmolinsky E, Karpasas M (2002) Characterization of the extracellular polysaccharide of Porphyridium sp.: molecular weight determination and rheological properties. Carbohydr Polym 50(2):183–189
Ghannoum MA, Rice LB (1999) Antifungal agents: mode of action, mechanisms of resistance, and correlation of these mechanisms with bacterial resistance. Clin Microbiol Rev 12(4):501–517
Ghasemi Y, Khalaj A, Mohagheghzadeh A, Khosravi AR, Morowvat MH (2007) Composition and antimicrobial activity of the essential oil and extract of Hypericum elongatum
Giammanco M et al (2015) Vitamin D in cancer chemoprevention. Pharm Biol 53:1399–1434
Gong M, Bassi A (2016) Carotenoids from microalgae: a review of recent developments. Biotechnol Adv 34:1396–1412
Gong X et al (2018) Carotenoid lutein selectively inhibits breast cancer cell growth and potentiates the effect of chemotherapeutic agents through ROS-mediated mechanisms. Molecules 23:905
Griffiths MJ, Harrison ST (2009) Lipid productivity as a key characteristic for choosing algal species for biodiesel production. J Appl Phycol 21:493–507
Guzmán S, Gato A, Lamela M, Freire-Garabal M, Calleja JM (2003) Anti-inflammatory and immunomodulatory activities of polysaccharide from Chlorella stigmatophora and Phaeodactylum tricornutum. Phytother Res 17(6):665–670
Hayakawa Y, Hayashi T, Lee JB, Srisomporn P, Maeda M, Ozawa T, Sakuragawa N (2000) Inhibition of thrombin by sulfated polysaccharides isolated from green algae. Biochim et Biophys Acta (BBA)-Protein Struct Mol Enzymol 1543(1):86–94

Hempel F, Maier UG (2016) Microalgae as solar-powered protein factories. Adv Exp Med Biol 896:241–262

Hempel F et al (2011) Microalgae as bioreactors for bioplastic production. Microb Cell Factories 10:1–6

Kang SM, Lee SH, Heo SJ, Kim KN, Jeon YJ (2011) Evaluation of antioxidant properties of a new compound, pyrogallol-phloroglucinol-6, 6'-bieckol isolated from brown algae, Ecklonia cava. Nutr Res Pract 5(6):495

Katiyar R, Arora A (2020) Health promoting functional lipids from microalgae pool: a review. Algal Res 46:101800

Khanra S et al (2018) Downstream processing of microalgae for pigments, protein and carbohydrate in industrial application: a review. Food Bioprod Process 110:60–84

Kim S-K, Kang K-H (2011) Medicinal effects of peptides from marine microalgae. Adv Food Nutr Res 64:313–323

Kim M, Yim JH, Kim SY, Kim HS, Lee WG, Kim SJ et al (2012) In vitro inhibition of influenza A virus infection by marine microalga-derived sulfated polysaccharide p-KG03. Antivir Res 93(2):253–259

Kiran BR, Venkata Mohan S (2021) Microalgal cell biofactory—therapeutic, nutraceutical and functional food applications. Plan Theory 10:836

Koller M et al (2014) Microalgae as versatile cellular factories for valued products. Algal Res 6:52–63

Kumar BR et al (2019) Microalgae as rich source of polyunsaturated fatty acids. Biocatal Agric Biotechnol 17:583–588

Kwak HS et al (2016) Synergistic effect of multiple stress conditions for improving microalgal lipid production. Algal Res 19:215–224

Laroche C, Michaud P (2007) New developments and prospective applications for β (1, 3) glucans. Recent Pat Biotechnol 1(1):59–73

Lee JB, Hayashi K, Hirata M, Kuroda E, Suzuki E, Kubo Y, Hayashi T (2006) Antiviral sulfated polysaccharide from Navicula directa, a diatom collected from deep-sea water in Toyama Bay. Biol. Pharm. Bull. 29:2135–2139

Levasseur W, Perré P, Pozzobon V (2020) A review of high value-added molecules production by microalgae in light of the classification. Biotechnol Adv 41:107545

Li Y et al (2008) Effects of nitrogen sources on cell growth and lipid accumulation of green alga Neochloris oleoabundans. Appl Microbiol Biotechnol 81:629–636

Li X et al (2020) Biotechnological production of astaxanthin from the microalga Haematococcus pluvialis. Biotechnol Adv 43:107602

Lin J-H et al (2015) Lutein production from biomass: Marigold flowers versus microalgae. Bioresour Technol 184:421–428

Liu H et al (2017) Protective effect of lutein on ARPE-19 cells upon H_2O_2-induced G_2/M arrest. Mol Med Rep 16:2069–2074

Long S et al (2018) Dietary supplementation with DHA-rich microalgae improves performance, serum composition, carcass trait, antioxidant status, and fatty acid profile of broilers. Poult Sci 97:1881–1890

Mata TM et al (2010) Microalgae for biodiesel production and other applications: a review. Renew Sust Energ Rev 14:217–232

Mishra A et al (2011) Characterization of extracellular polymeric substances produced by microalgae Dunaliella salina. Carbohydr Polym 83:852–857

Mostafa SS (2012) Microalgal biotechnology: prospects and applications. Plant Sci 12:276–314

Mourelle ML et al (2017) The potential use of marine microalgae and cyanobacteria in cosmetics and thalassotherapy. Cosmetics 4:46

Mudimu O, Rybalka N, Bauersachs T, Born J, Friedl T, Schulz R (2014) Biotechnological screening of microalgal and cyanobacterial strains for biogas production and antibacterial and antifungal effects. Metabolites 4(2):373–393

Najdenski HM, Gigova LG, Iliev II, Pilarski PS, Lukavský J, Tsvetkova IV et al (2013) Antibacterial and antifungal activities of selected microalgae and cyanobacteria. Int J Food Sci Technol 48(7):1533–1540

Nauroth JM, Liu YC, Van Elswyk M, Bell R, Hall EB, Chung G, Arterburn LM (2010) Docosahexaenoic acid (DHA) and docosapentaenoic acid (DPAn-6) algal oils reduce inflammatory mediators in human peripheral mononuclear cells in vitro and paw edema in vivo. Lipids 45:375–384

Odjadjare EC et al (2017) Potential biotechnological application of microalgae: a critical review. Crit Rev Biotechnol 37:37–52

Ohgami K, Shiratori K, Kotake S, Nishida T, Mizuki N, Yazawa K, Ohno S (2003) Effects of astaxanthin on lipopolysaccharide-induced inflammation in vitro and in vivo. Invest Ophthalmol Vis Sci 44(6):2694–2701

Olasehinde TA et al (2017) Therapeutic potentials of microalgae in the treatment of Alzheimer's disease. Molecules 22:480

Park JS, Chyun JH, Kim YK, Line LL, Chew BP (2010) Astaxanthin decreased oxidative stress and inflammation and enhanced immune response in humans. Nutr Metab 7:1–10

Pereira H et al (2018) Fluorescence activated cell-sorting principles and applications in microalgal biotechnology. Algal Res 30:113–120

Posser C et al (2018) Supplementation of diets with omega-3 fatty acids from microalgae: effects on sow reproductive performance and metabolic parameters. Livest Sci 207:59–62

Prakash JW, Marimuthu J, Jeeva S (2011) Antimicrobial activity of certain fresh water microalgae from Thamirabarani River, Tamil Nadu, South India. Asian Pac J Trop Biomed 1(2):S170–S173

Ranadheer P, Kona R, Sreeharsha RV, Mohan SV (2019) Non-lethal nitrate supplementation enhances photosystem II efficiency in mixotrophic microalgae towards the synthesis of proteins and lipids. Bioresour Technol 283:373–377

Raposo MFDJ, Oliveira SE, Castro PM, Bandarra NM, Morais RM (2010) On the utilization of microalgae for brewery effluent treatment and possible applications of the produced biomass. J. Inst. Brew. 116:285–292

Rasmussen HM, Johnson EJ (2013) Nutrients for the aging eye. Clin Interv Aging 8:741–748

Rendueles O, Ghigo JM (2012) multi-species biofilms: How to avoid unfriendly neighbours. FEMS Microbiol. Rev. 36:972–989

Rodríguez-Meizoso I, Jaime L, Santoyo S, Señoráns FJ, Cifuentes A, Ibáñez E (2010) Subcritical water extraction and characterization of bioactive compounds from Haematococcus pluvialis microalga. J Pharm Biomed Anal 51(2):456–463

Rohit M, Venkata Mohan S (2018) Quantum yield and fatty acid profile variations with nutritional mode during microalgae cultivation. Front Bioeng Biotechnol 6:111

Sabeela Beevi U, Sukumaran RK (2015) Cultivation of the fresh water microalga *Chlorococcum* sp. RAP13 in sea water for producing oil suitable for biodiesel. J Appl Phycol 27:141–147

Samarakoon KW, O-Nam K, Ko JY, Lee JH, Kang MC, Kim D et al (2013) Purification and identification of novel angiotensin-I converting enzyme (ACE) inhibitory peptides from cultured marine microalgae (Nannochloropsis oculata) protein hydrolysate. J Appl Phycol 25:1595–1606

Sampathkumar SJ, Gothandam KM (2019) Sodium bicarbonate augmentation enhances lutein biosynthesis in green microalgae Chlorella pyrenoidosa. Biocatal Agric Biotechnol 22:101406

Sansone C et al (2017) The green microalga Tetraselmis suecica reduces oxidative stress and induces repairing mechanisms in human cells. Sci Rep 7:41215

Santoyo S, Plaza M, Jaime L, Ibañez E, Reglero G, Señorans J (2011) Pressurized liquids as an alternative green process to extract antiviral agents from the edible seaweed Himanthalia elongata. J Appl Phycol 23:909–917

Sathasivam R et al (2019) Microalgae metabolites: a rich source for food and medicine. Saudi J Biol Sci 26:709–722

Scott SA et al (2010) Biodiesel from algae: challenges and prospects. Curr Opin Biotechnol 21:277–286

Senni K, Pereira J, Gueniche F, Delbarre-Ladrat C, Sinquin C, Ratiskol J et al (2011) Marine polysaccharides: a source of bioactive molecules for cell therapy and tissue engineering. Mar Drugs 9(9):1664–1681

Shanab SM, Mostafa SS, Shalaby EA, Mahmoud GI (2012) Aqueous extracts of microalgae exhibit antioxidant and anticancer activities. Asian Pac J Trop Biomed 2(8):608–615

Sheih IC, Fang TJ, Wu TK, Lin PH (2010) Anticancer and antioxidant activities of the peptide fraction from algae protein waste. Journal of agricultural and food chemistry. 58(2):1202–1207

Silva ADS, de Magalhães WT, Moreira LM, Rocha MVP, Bastos AKP (2018) Microwave-assisted extraction of polysaccharides from Arthrospira (Spirulina) platensis using the concept of green chemistry. Algal Res 35:178–184

Singh G, Beddow J, Mee C, Maryniak L, Joyce EM, Mason TJ (2017) Cytotoxicity study of textile fabrics impregnated with CuO nanoparticles in mammalian cells. Int J Toxicol 36(6):478–484

Sreeharsha RV, Mohan SV (2020) Obscure yet promising oleaginous yeasts for fuel and chemical production. Trends Biotechnol 38:873–887. https://doi.org/10.1016/j.tibtech.2020.02.004

Sreeharsha RV, Venkata Mohan S (2021) Symbiotic integration of bioprocesses to design a self-sustainable life supporting ecosystem in a circular economy framework. Bioresour Technol 326:124712

Sreeharsha RV, Tejaswini GS, Mohan SV (2022) Dual-stage biorefinery to convert spentwash hydrolysate into oleochemicals using Trichosporon cutaneum and Yarrowia lipolytica. Bioresour Technol 354:127146

Sreeharsha RV et al (2023) Orienting biodiesel production towards sustainability and circularity by tailoring the feedstocks and processes. J Clean Prod 414:137526

Sun L, Wang L, Zhou Y (2012) Immunomodulation and antitumor activities of different-molecular-weight polysaccharides from Porphyridium cruentum. Carbohydr Polym 87(2):1206–1210

Suparmi S et al (2016) Anti-anemia effect of chlorophyll from katuk (Sauropus androgynus) leaves on female mice induced sodium nitrite. Pharmacogn J 8:375–379

Talukdar J, et al. (2020) COVID-19: potential of microalgae derived natural astaxanthin as adjunctive supplement in alleviating cytokine storm. https://doi.org/10.2139/ssrn.3579738

Talyshinsky MM, Souprun YY, Huleihel MM (2002) Anti-viral activity of red microalgal polysaccharides against retroviruses. Cancer Cell Int 2:1–7

Vijayakumar S, Menakha M (2015) Pharmaceutical applications of cyanobacteria—a review. JACME 5:15–23

Vo TS, Ngo DH, Van Ta Q, Kim SK (2011) Marine organisms as a therapeutic source against herpes simplex virus infection. European Journal of Pharmaceutical Sciences. 44(1-2):11–20

Wang H, Liu Y, Gao X, Carter CL, Liu ZR (2007) The recombinant β subunit of C-phycocyanin inhibits cell proliferation and induces apoptosis. Cancer Lett 247(1):150–158

Yeh K-L, Chang J-S (2012) Effects of cultivation conditions and media composition on cell growth and lipid productivity of indigenous microalga Chlorella vulgaris ESP-31. Bioresour Technol 105:120–127

Zhao CS, Shao NF, Yang ST, Ren H, Ge YR, Feng P et al (2019) Predicting cyanobacteria bloom occurrence in lakes and reservoirs before blooms occur. Sci Total Environ 670:837–848

Chapter 7
Bacterial Photosynthesis for Nutrient Recovery and Value Addition

Abstract Cyanobacteria have a high yield potential per-acre and are adaptable to environmental changes and can grow in different types of water resources. These organisms are extensively employed as model organisms in the field of research to investigate photosynthetic mechanisms and diverse responses to abiotic stress. Cyanobacteria play a significant role in providing a diverse range of primary and secondary metabolites that are utilized in many applications such as biofuels, biochemical synthesis, pharmaceuticals, biopesticides, therapeutics, coloring pigments, and food additives. Cyanobacteria also have the potential to sequester carbon dioxide and produce biohydrogen as an ideal energy carrier for future generations. Cyanobacteria produce PHAs directly from sunlight and CO_2, with the dominant type being polyhydroxybutyrate. Cyanobacteria possess the ability to harness solar energy and produce electrical energy via light-dependent electrogenic activity in microbial fuel cells (MFCs). Cyanobacteria have gained recognition as a significant microorganism in the context of sustainable agricultural advancement, owing to their capacity to breakdown diverse contaminants and fulfill several functions within the soil ecosystem. Diazotrophes are a type of cyanobacteria that can generate eco-friendly biofertilizers, which are easily available and less costly than traditional chemical-based fertilizers. Cyanobacteria also play a critical role in maintaining and building soil fertility by making soil porous, improving water-holding capacity, controlling weeds growth, and increasing the availability of soil phosphate.

Keywords Cyanobacteria · Bioelectricity · Biofertilizers · Polyhydroxybutyrate · Microbial fuel cells · Biofuels

7.1 Introduction

Cyanobacteria are facultative aerobic photosynthetic microorganisms belonging to the phylum Cyanophyta. The RUBISCO enzyme of cyanobacteria is the most evolved and conserved enzyme that functions in the environment where the CO_2 levels are

R. V. Sreeharsha, S. Venkata Mohan, *Microbial Photosynthesis*,
https://doi.org/10.1007/978-981-97-1253-3_7

several folds higher than the current levels. Due to their primitive nature and evolutionary conservation, cyanobacteria can be considered model organisms for exploring photosynthesis, circadian rhythms, and nitrogen fixation. Further, the great extent of similarity of oxygenic photosynthesis and related metabolic pathways with that of higher plants enables scientists to use these unicellular organisms to deeply understand the obscure photosystem mechanisms. In Chap. 1, we discussed about fundamental characteristics of cyanobacteria and their taxonomical status in oxygenic photosynthesis. Cyanobacteria have the capacity to thrive in various water sources, including freshwater, brackish water, and industrial wastewater. Their advantages spread beyond carbon fixation and oxygen evolution. Scientists have discovered their enormous potential as industrial workhorses or cell factories to produce value-added compounds. The ability to grow in diverse habitats and to adapt to different stress conditions made them more dependable organisms for industrial applications. When compared to higher plants that fix carbon and produce useful compounds, cyanobacteria are superior resources as they require low land requirements and fewer resources.

The carbon fixation through cyanobacteria generates nutrient-rich biomass which can be processed through various techniques into diverse compounds. The advancements in harvesting techniques, biomass processing, and pretreatment favor the cyanobacteria-based biorefinery. The range of compounds that can be generated from cyanobacteria depends on the growth conditions, nutrient feed, and culture conditions. The primary and secondary metabolites produced through the complex metabolic pathways of cyanobacteria are the resources for biofuels and biochemical production and also active ingredients for pharmaceutical, biopesticide, and therapeutic applications. Moreover, they serve as a significant reservoir of bioactive substances, biopolymers, coloring agents, dietary supplements, and biofertilizers. Cyanobacteria have the capacity to synthesize a diverse array of secondary metabolites, which encompass alkaloids, terpenoids, shikimate-derived chemicals, amino glycosides, as well as peptide groups of compounds such as polyketides or nonribosomal peptides.

The cellular structure and biochemical composition of cyanobacteria revealed the presence of specialized structures or aggregates which are called inclusion bodies. These inclusion bodies are composed of various substances and play different roles within the cells. They help cyanobacteria adapt to changing environmental conditions and ensure their survival during periods of nutrient scarcity or stress. Each type of inclusion body has a specific function and contributes to the overall metabolic flexibility and resilience of cyanobacteria. The well-studied inclusion bodies in cyanobacteria include carboxysomes, gas vesicles, polyphosphate granules, glycogen granules, cyanophycin, and polyhydroxybutyrate granules. Carboxysomes are composed of protein shells and contain the enzyme RuBisCO, which is essential for carbon fixation during photosynthesis. Carboxysomes compartmentalize the process of carbon fixation, allowing cyanobacteria to efficiently concentrate carbon dioxide around RuBisCO, thereby enhancing photosynthetic efficiency. Hollow gas vesicles and gas-filled structures provide buoyancy to cyanobacterial cells, allowing them to regulate their position in the water column. They provide the cells with the ability to move toward favorable light conditions or nutrient concentrations. Polyphosphate granules, volutin granules, or metachromatic granules store excess phosphorus and serve as a reserve of phosphorus that can be utilized during times of nutrient limitation. Glycogen granules can accumulate glycogen as a storage polysaccharide and act as an energy reserve and can be broken down to provide the cell with glucose for energy production during periods of low light or nutrient availability. Cyanophycin is a nitrogen-rich polymer synthesized by cyanobacteria during periods of nitrogen excess and serves as a reserve of nitrogen. It can be stored in cyanophycin granules and can supply the cell with amino acids when nitrogen availability is limited. Polyhydroxybutyrate (PHB) is a type of storage polymer that is stored in PHB granules and serves as a carbon and energy reserve and can be used by the cells during periods of carbon limitation. These inclusion bodies which play a crucial role in cyanobacteria growth and metabolism were alternatively explored for producing value-added compounds.

7.2 Diverse Groups of Products Obtained from Cyanobacteria

7.2.1 Biodiesel

The use of fossil fuels in the transportation sector raises concerns about greenhouse gas emissions and contributes to the increase in Earth's temperature. Further, fossil fuels are exhaustible and hence cannot be a reliable option for future energy needs. Physical and chemical techniques including absorption, adsorption, cryogenic, and membrane separation are useful in CO_2 capture, nevertheless, biological routes of CO_2 fixation are always reliable and sustainable. On the other hand, biofuels that are derived from renewable resources could be the sustainable alternatives for petro-based fuels and help in the reduction of emissions (Sreeharsha and Mohan 2020; Sreeharsha et al. 2023; Kumar et al. 2017). Cyanobacteria are an ideal source for carbon sequestration coupled with biofuel production (Farrokh et al. 2019). They possess the unique ability to withstand high concentrations of CO_2 and are optimal to grow in localities where the CO_2 concentrations are unusually high such as power plants, flue gas, and mining sites (Parmar et al. 2011). Cyanobacteria have significant metabolic flexibility, since they have the ability to ingest not only carbon dioxide (CO_2) and oxygen (O_2), but also nitrogen gas (N_2) and hydrogen gas (H_2) (Mund et al. 2022).

7.2.2 Biohydrogen

The research findings during the last decade have proven that biohydrogen could be the future of the energy sector due to its characteristic features like high heating value and low CO_2 emissions (Khetkorn et al. 2023). In cyanobacteria, the molecular hydrogen that is produced out of water hydrolysis and electrons from reduced ferredoxin will be converted into hydrogen by a hydrogenase catalyzed reaction. In addition to light-dependent ferredoxin reduction, the cyanobacteria can perform dark fermentation for hydrogen production in which pyruvate forms the central hub for electron generation through oxidoreductase reactions. Several cyanobacteria species including *Calothrix elenkinii*, *Fischerella muscicola*, *Nostoc calcicola*, *Scytonema bohneri*, and *Tolypothrix distorta* are known for their efficacy for biohydrogen production through photolysis as well as dark fermentation routes (Bozieva et al. 2023; Kumar et al. 2019). It was evident from the studies on *F. muscicola* and *Anabaena* UTEX 1448 that biohydrogen production can be double enhanced in the nitrogen- and sulfur-deprived conditions when compared to control conditions (Vargas et al. 2018; Yodsang et al. 2018).

7.2.3 Polyhydroxyalkanoates

Natural plastics mostly originate from non-renewable petrochemical sources and have limited biodegradability, hence giving rise to significant environmental concerns, particularly with regard to human health. In recent times, there has been a notable surge in the interest of scholars toward the substitution of petrochemical-based plastics with biodegradable polymers, commonly referred to as bioplastics. These bioplastics are manufactured using sustainable resources including vegetable oils, fats, maize starch, and microbes. As mentioned earlier, PHAs are polymeric storage substances produced by cyanobacteria through natural metabolic pathways usually under nitrogen and phosphate depletion (Kaewbai-ngam et al. 2022; Wang et al. 2021). Among other PHAs, polyhydroxybutyrate (PHB) is popularly studied for its hydrophobicity, thermoplastic processability, and diverse applications. The compound is derived through the process of polymerization involving hydroxybutyryl CoA monomers. These monomers are generated from acetyl CoA through acetylation and reduction by β-ketothiolase and acetoacetyl CoA reductase enzymes, respectively. In general, the acetyl CoA is obtained from the degradation of long-chain fatty acids (C_{16} and C_{18}) through β-oxidation pathway. Alternatively, the PHA synthase enzyme catalyzes the polymerization of hydroxyacyl CoA thioester monomers (Lane and Benton 2015). The accumulation of PHB can be regarded as a buffering mechanism employed to manage surplus acetyl CoA and NADH. Under conditions of darkness and stressed fermentation, polyhydroxybutyrate (PHB) serves as the secondary storage component, following glycogen, that undergoes degradation for the purpose of energy production in the tricarboxylic acid (TCA) cycle. Cyanobacteria classified within the taxonomic order Nostocales exhibit the greatest capacity for polyhydroxyalkanoate (PHA) accumulation when subjected to photoautotrophic and mixotrophic conditions, resulting in PHA content reaching as high as 78% of the total dry cell weight (Gomes Gradíssimo et al. 2020). Genera including *Leptolyngbya*, *Calothrix*, *Chlorogloea*, *Spirulina*, *Aphanothece*, *Gloeothece*, and *Synechococcus* are also known to produce high quantities of

PHA. The production of polyhydroxybutyrate (PHB) and other polyhydroxyalkanoates (PHAs) by cyanobacteria is rather modest, ranging from 0.5% to 6% of the dry mass (equivalent to approximately 13 mg/L/h). In contrast, heterotrophic bacteria exhibit significantly higher production rates of 3.2 g/L/h. Moreover, the production process necessitates the presence of nutrient insufficiency, together with the incorporation of specific organic carbon sources.

Despite the slow production rate, wild-type cyanobacteria are capable of accumulating considerable fractions of carbon in the form of PHA under photoautotrophic conditions. Some examples of this are *Spirulina platensis* (6% DCW of PHB), *Gloeothece* sp. (6% DCW of PHB), *Synechococcus* sp. MA19 (~30% DCW of PHB), and *Oscillatoria limosa* (6% DCW of poly-*β*-hydroxyvalerate). Significant fractions of PHB accumulation have been reached in *Syn*PCC6803 (~38% DCW), *Nostoc muscorum* (~78% DCW), and *Aulosira fertilissima* (~85% DCW) by modifying the culture conditions and making use of stressful conditions such as phosphate and nitrogen starvation and growth in mixotrophic metabolism by using acetate, citrate, and other sugars (Drosg et al. 2015). Nevertheless, the real benefits of applying cyanobacteria to produce bioplastics or PHAs will come from photoautotrophic metabolism as it disregards usage of inorganic carbon. To achieve this, synthetic biology approaches that modulate the metabolic flux to produce PHA in photosynthetic routes will be highly useful. Efforts have been directed toward enhancing the inherent PHB biosynthesis pathway in cyanobacteria through metabolic modifications, with a specific emphasis on including enzymes from other bacteria capable of generating PHA (Wang et al. 2014).

Synechocystis sp. PCC 6803 was engineered to produce PHB at 25% through genome editing and through transposon mutagenesis and expressing genes from other organisms. Several genetic modifications have been conducted on the model cyanobacterium, Synechocystis, in order to augment the production of several biological substances, including ethanol, hydrogen, glycerol, and polyhydroxyalkanoates (PHAs) (Carpine et al. 2017). The genetic engineering technique of overexpressing the sigma factor sigE in Synechocystis sp. PCC 6803 resulted in alterations to the sugar catabolism pathway, ultimately resulting to an enhanced synthesis of polyhydroxybutyrate (PHB) under conditions of nitrogen limitation. (Osanai et al. 2013). It is worth noting that the overexpression of PHA synthesis genes in numerous cyanobacterial species has led to a modest elevation in PHA levels. However, in order to enhance PHA production in other cyanobacterial species, more intricate metabolic modifications are required.

The biodegradability and biocompatibility of PHB enable it to be used as bioplastics, medical and pharmaceutical compounds, cosmetics and personal care products, and biofuels. They found applications in packaging, disposable cutlery, and other single-use items, agriculture, environmental remediation, drug delivery systems, sutures, tissue engineering scaffolds, biodegradable textiles, fibers, and non-woven materials. They are especially useful in controlled release of fertilizers and pesticides and improve soil quality through PHB-based mulches (Balaji et al. 2013).

7.2.4 Exopolysaccharides

Exopolysaccharides are another class of useful compounds with biotechnological applications produced from cyanobacteria on an industrial scale. EPSs are composed of a minimum of 10 distinct monosaccharides and serve as the primary reservoir of pentoses, which are typically lacking in other EPS derived from prokaryotes. Important EPS-producing cyanobacteria include halophilic *Aphanocapsa halophytica*, *Anabaena* sp. ATCC 33047, and *Cyanothece* sp. ATCC 51142 (Matsunaga et al. 1996; Moreno et al. 1998). EPSs found applications in the biomedical industry and are also useful in oil recovery and food emulsification (Otero and Vincenzini 2003). It is deemed advantageous to conduct screening and isolation procedures for cyanobacterial strains that exhibit enhanced capacity for extracellular polymeric substance (EPS) synthesis. Cyanobacteria synthesize EPS through extracellular sugars or glucose-6-phosphate as a precursor with the involvement of glycosyltransferase enzymes. EPS helps cyanobacteria in protection from desiccation, UV radiation, and physical damage and for attachment to surfaces, aiding in biofilm formation as well as for nutrient trapping (Bhunia et al. 2018). During growth, EPS promotes cell aggregation, leading to the formation of flocs that sediment from the water column.

7.2.5 Bioelectricity

Microbial fuel cells (MFCs) are specialized bioreactors that leverage microorganisms as biocatalysts to convert chemical energy into electrical energy. The configuration of the MFC comprises two distinct chambers, namely the anode and the cathode, which are physically separated by a proton exchange membrane (Sjöblom et al. 2017). One significant challenge faced by MFCs is the high cost associated with the PEM (Jang et al. 2004). The process of electron transfer from the surfaces of non-conductive microbial cells to the electrode in microbial electron transport systems presents an additional challenge (Wei et al. 2011). In order to tackle this issue, electrochemical mediators are utilized to assist the flow of electrons from microbial cells to the electrode. Nevertheless, these mediators possess characteristics that are detrimental to health and incur substantial expenses. A potentially advantageous alternative methodology has surfaced, which entails the utilization of microbial fuel cells (MFCs) that obviate the requirement for membranes and electrochemical mediators. This development significantly improves the cost feasibility of wastewater treatment (Jang et al. 2004).

Researchers have undertaken an innovative approach by investigating the utilization of photosynthetic microorganisms with the ability to harness solar energy, as opposed to depending solely on heterotrophic bacteria. One noteworthy experiment was the utilization of a microbial fuel cell (MFC) that housed a monoculture of the cyanobacterium Spirulina platensis. This MFC functioned in the absence of electrochemical mediators and predominantly produced electricity via heterotrophic metabolism. Meanwhile, photosynthesis principally facilitated the synthesis of glycogen (Fu et al. 2009). Further investigations indicated that cyanobacteria could transmit electrons to their surroundings under illumination, showcasing their potential for light-dependent electrogenic activity (Pisciotta et al. 2010). This process was mediated by plastoquinone (PQ), with photosystem (PS) I and PS II inhibition. Usually, electron generation is observed in the presence of light during photosynthetic oxygenation, but glycogen breakdown primarily occurs in the absence of light, assisted by the mediator 2-hydroxy-1,4-naphthoquinone.

Under stress conditions, thermophilic microorganisms exhibit distinct physiological responses. Ko et al. (2015) demonstrated that the thermophilic cyanobacterium TCL-1, originating from the Chin-Lun hot spring at 62 °C, exhibited a photocurrent response at 0.4 V in an H-type MFC without the need for electrochemical mediators, with electron outflow originating from the inner cells. Furthermore, it was observed by Ko et al. (2015) that TCL-1 cells adhered to the anodic membrane exhibited a greater lipid content compared to cells that were in suspension. In a recent advancement, researchers utilized the identical thermophilic cyanobacterial strain to produce photocurrent without the incorporation of artificial mediators. The present investigation successfully identified plastoquinone (PQ), cytochrome b6/f, and plastocyanin as electron transport mediators within a voltage range of 0.2 to 0.7 V (Ko et al. 2018). A wide range of value-added compounds that are obtained from cyanobacterial species are represented in (Fig. 7.1 and Table 7.1).

7.2.6 Biofertilizers

The persistently growing population especially in south Asian countries demands contamination-free and healthy food. In the last few decades, green revolution practices significantly contributed to the increase in food production. However, the use of chemical fertilizers had serious challenges in terms of chemical effluents in the environment and toxic remnants in food products. The use of biofertilizers which are derived from natural sources such as cyanobacteria increases soil fertility and crop productivity leading to an eco-friendly environment and sustainable agricultural development. Also, the use of biofertilizers derived from diazotrophes cyanobacteria can mitigate a wide range of toxic elements in the soil ecosystem and help in weed control (Singh et al. 2017). Atmospheric nitrogen fixation is another important phenomenon through which cyanobacteria contribute to soil aeration and enrichment. Cyanobacteria species such as *Anabaena variabilis* and *Nostoc linckia*, *Aulosira fertilissima*, *Calothrix* sp.,

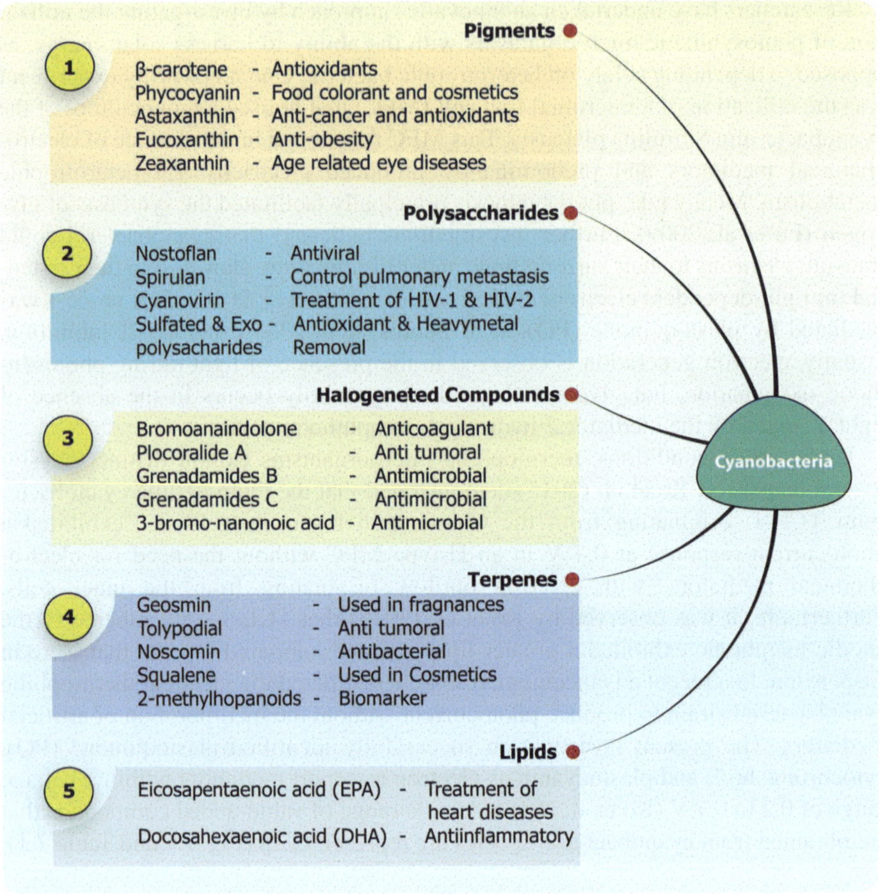

Fig. 7.1 Diverse product portfolio obtained from cyanobacteria and their applications

Tolypothrix sp., and *Scytonema* sp., colonize on soil and rocks and fix up to 20–25 kg/ha atmospheric nitrogen. Cyanobacteria produce phytohormones, vitamins (especially vitamin B12, which is known to enhance water-holding capacity), phosphates, and amino acids in soil, thus enriching the soil properties (Song et al. 2005). Cyanobacterial-derived biofertilizers have shown promising results in barley, oats, tomato, radish, cotton, sugarcane, maize, chilli, and lettuce (Saadatnia and Riahi 2009; Wilson et al. 2006; Ibraheem 2007).

The formulation of biofertilizer is unique wherein it is a combination of inoculum and carrier. Carriers are usually inert compounds with good moisture absorption capacity and are easy to sterilize with excellent buffering capacity. The carrier material exhibits convenient handling properties, an extended shelf life, and enhances the efficacy of the biofertilizer. The carrier such as peat will be effective

Table 7.1 Summary of different cyanobacterial species and the corresponding value-added compounds produced

Cyanobacterial species	Targeted product	Cultivation mode	References
Bioplastics			
Synechocystis sp.	Poly 3-hydroxybutyrate (PHB)	Mixotrophic with N2 limited condition	Osanai et al. (2013)
Aulosira fertilissima	Poly(3 hydroxybutyrate-co-3-hydroxyvalerate)	Heterotrophic with phosphate deficiency	Singh et al. (2016)
Spirulina platensis	Poly 3-hydroxybutyrate (PHB)	Photoautotrophic	Singh et al. (2016)
Biofertilizers			
Nostoc commune	Biofertilizer	Mixotrophic	Chittapun et al. (2020)
Anabaena azollae	Nitrogen supplement	–	Reddy and Triveni
Tolypothrix tenuis	Nitrogen fixation	–	Singh et al. (2016)
Biofuels			
Synechococcus elongatus	2,3 Butanediol		Oliver et al. (2013)
Synechocystis sp.	Ethanol	Mixotrophic	Gao et al. (2012)
Anabaena cylindrica	Biohydrogen	Mixotrophic	Masukawa e al. (2002)
Biotherapeutics			
Lyngbya majuscula	Dragomabin	Flask culture	McPhail et al. (2007)
Anabaena circinalis	Anatoxin-a	Mixotrophic	Jha and Zi-Rong (2004)
Gracilaria dendroides	Kaempferol	Flask culture	Al-Saif et al. (2014)

in combining with approximately 10^8 powdered or dried viable cells capacity at a pH range of 6.0 to 7.5 and temperature between 25 and 30 °C. Additional substances that have demonstrated efficacy as transporters encompass lignite, coal, charcoal, filter mud, vermiculite, polyacrylamide, mineral soils, vegetable oils, coir waste, and clay. Bacterial inoculants including rhizobia, nitrogen-fixing rhizobacteria, plant growth-promoting rhizobacteria, and phosphate-solubilizing bacteria are compatible with carrier materials and act as effective biofertilizers. Powder formulations based on carrier elements are convenient to handle, package, and store and suitable for soil application. Alternatively, dried cyanobacterial biomass can be prepared into granules by combining the biomass with binders and solid carriers. Granules offer a controlled release of nutrients and are particularly useful for slow-release applications. However, the disadvantages associated with the carrier are its processing through mining, drying, and milling which is expensive to make the pore size 10–40 μm.

The formulation of biofertilizers from cyanobacteria involves preparing the cyanobacterial biomass in a way that it can be easily applied to soil or plants while maintaining the viability and effectiveness of the beneficial microorganisms. The formulation process can vary based on the desired product type, storage stability, and application method. The harvested cyanobacterial biomass is dewatered through mechanical pressing, filtration, or centrifugation and dried to further reduce moisture content. Ensuring that the stability and shelf life of the biomass are of utmost importance, making this stage critical. The cyanobacterial biomass can be suspended in either water or a nutrient solution when formulating liquid biofertilizers. This type of formulation is suitable for foliar application and can also be applied through irrigation systems. Properties of biofertilizers can be enhanced by adding certain additives such as stabilizers, growth-promoting substances, and nutrients to further improve stability, nutrient content, and shelf life. Upon successful synthesis of biofertilizer, the effectiveness shall be tested by assessing the number of viable cells per unit of the product. Subsequently, the formulated biofertilizer is packaged in appropriate containers that protect from moisture, light, and external contaminants. Storage conditions such as temperature and humidity need to be carefully controlled to maintain the viability of the microorganisms. The formulated biofertilizer can be applied to soil, plants, or seeds using various application methods, including broadcasting, seed coating, mixing into soil, foliar spraying, and drip irrigation. After application, regular monitoring of soil nutrient levels and plant growth should be conducted to assess the effectiveness of the biofertilizer and make any necessary adjustments to application rates or schedules.

7.3 Conclusion

Cyanobacteria hold a great promise in the fields of biotechnology, renewable energy, and agriculture due to their unique abilities and metabolic pathways. They act as an ideal feedstock for producing biodiesel, biohydrogen, polyhydroxyalkanoates, exopolysaccharides, biofertilizers, and bioelectricity. The ability in utilizing wastewaters to produce value-added products made them unique in biorefinery applications. The formulation of these products involves specific processes, such as biomass harvesting, drying, and formulation with suitable carriers or additives to enhance stability and effectiveness. Overall, cyanobacteria offer sustainable solutions for a range of industrial and agricultural needs, and ongoing research is likely to uncover even more applications and optimize production methods for these valuable products.

References

Al-Saif SSAL, Abdel-Raouf N, El-Wazanani HA, Aref IA (2014) Antibacterial substances from marine algae isolated from Jeddah coast of Red sea, Saudi Arabia. Saudi J Biol Sci 21(1):57–64

Balaji B, Xu J, Nwokafor A, Gupta R, Agarwal Y (2013) Sentinel: occupancy based HVAC actuation using existing WiFi infrastructure within commercial buildings. In: Proceedings of the 11th ACM conference on embedded networked sensor systems, pp 1–14

Bhunia B, Uday USP, Oinam G, Mondal A, Bandyopadhyay TK, Tiwari ON (2018) Characterization, genetic regulation and production of cyanobacterial exopolysaccharides and its applicability for heavy metal removal. Carbohydr Polym 179:228–243

Bozieva AM et al (2023) New cyanobacterial strains for biohydrogen production. Int J Hydrog Energy 48:7569–7581

Carpine R, Du W, Olivieri G, Pollio A, Hellingwerf KJ, Marzocchella A, dos Santos FB (2017) Genetic engineering of Synechocystis sp. PCC6803 for poly-β-hydroxybutyrate overproduction. Algal Res 25:117–127

Chittapun S, Jonjaroen V, Khumrangsee K, Charoenrat T (2020) C-phycocyanin extraction from two freshwater cyanobacteria by freeze thaw and pulsed electric field techniques to improve extraction efficiency and purity. Algal Res 46:101789

Drosg B et al (2015) Photo-autotrophic production of poly (hydroxyalkanoates) in cyanobacteria. Chem Biochem Eng Q 29:145–156

Farrokh P et al (2019) Cyanobacteria as an eco-friendly resource for biofuel production: a critical review. Biotechnol Prog 35:e2835

Fu CC, Su CH, Hung TC, Hsieh CH, Suryani D, Wu WT (2009) Effects of biomass weight and light intensity on the performance of photosynthetic microbial fuel cells with Spirulina platensis. Bioresour Technol 100(18):4183–4186

Gao Q, Wang W, Zhao H, Lu X (2012) Effects of fatty acid activation on photosynthetic production of fatty acid-based biofuels in Synechocystis sp. PCC6803. Biotechnol Biofuels 5:1–9

Gomes Gradíssimo D et al (2020) Cyanobacterial polyhydroxyalkanoates: a sustainable alternative in circular economy. Molecules 25:4331

Ibraheem I (2007) Cyanobacteria as alternative biological conditioners for bioremediation of barren soil. Egypt J Phycol 8(1):99–117

Jha RK, Zi-Rong X (2004) Biomedical compounds from marine organisms. Marine drugs. 2(3):123–146

Jang JK, Pham TH, Chang IS, Kang KH, Moon H, Cho KS, Kim BH (2004) Construction and operation of a novel mediator-and membrane-less microbial fuel cell. Process Biochem 39(8):1007–1012

Kaewbai-ngam J et al (2022) Production of glycogen, PHB, biohydrogen, NAD (P) H, and proteins in Synechocystis sp. PCC 6803 disrupted in metabolically linked biosynthetic pathway (s). J Appl Phycol 34:1983–1995

Khetkorn W et al (2023) Photobiohydrogen production and in cyanobacteria strategies for H_2 yield improvements in cyanobacteria. Adv Biochem Eng Biotechnol 183:253–280

Ko TH, Hsu YC, Wang YF, Chang YH, Lin SK, Hsueh HT, Chen HH (2015) Variations of cellular components in Thermosynechococcus sp. CL-1 under electrochemical treatment. RSC advances. 5(84):68470–68475

Ko TH, Chang FH, Hsueh HT, Wang YF, Su CM, Huang LC, Chu H, Tseng CM, Ray DT, Shen YH (2018) Photocurrent production with thermophilic cyanobacterial strain under electrochemical treatment without adding of mediators. International Journal of Electrochemical Science. 13(4):3429–3440

Kumar S, Sreeharsha RV, Mudalkar S, Sarashetti PM, Reddy AR (2017) Molecular insights into photosynthesis and carbohydrate metabolism in Jatropha curcas grown under elevated CO2 using transcriptome sequencing and assembly. Sci Rep 7(1):11066

Kumar S et al (2019) Bioprospecting of microbes for biohydrogen production: current status and future challenges. In: Bioprocessing for biomolecules production. Wiley, Hoboken, pp 443–471

Lane CE, Benton MG (2015) Detection of the enzymatically-active polyhydroxyalkanoate synthase subunit gene, phaC, in cyanobacteria via colony PCR. Mol Cell Probes 29:454–460

Masukawa H, Mochimaru M, Sakurai H (2002) Disruption of the uptake hydrogenase gene, but not of the bidirectional hydrogenase gene, leads to enhanced photobiological hydrogen production by the nitrogen-fixing cyanobacterium Anabaena sp. PCC 7120. Appl Microbiol Biotechnol 58:618–624

Matsunaga T, Park CH, Bessho T, Mu D, Sancar A (1996) Replication protein A confers structure-specific endonuclease activities to the XPF-ERCC1 and XPG subunits of human DNA repair excision nuclease (*). J Biol Chem 271(19):11047–11050

McPhail KL, Correa J, Linington RG, González J, Ortega-Barría E, Capson TL, Gerwick WH (2007) Antimalarial linear lipopeptides from a Panamanian strain of the marine cyanobacterium Lyngbya majuscula. J Nat Prod 70(6):984–988

Moreno J, Vargas MA, Olivares H, Rivas J, Guerrero MG (1998) Exopolysaccharide production by the cyanobacterium Anabaena sp. ATCC 33047 in batch and continuous culture. J Biotechnol 60(3):175–182

Mund NK et al (2022) Advances in metabolic engineering of cyanobacteria for production of biofuels. Fuel 322:124117

Oliver JW, Machado IM, Yoneda H, Atsumi S (2013) Cyanobacterial conversion of carbon dioxide to 2, 3-butanediol. Proceedings of the National Academy of Sciences. 110(4):1249–1254

Osanai T, Numata K, Oikawa A, Kuwahara A, Iijima H, Doi Y et al (2013) Increased bioplastic production with an RNA polymerase sigma factor SigE during nitrogen starvation in Synechocystis sp. PCC 6803. DNA Res 20(6):525–535

Otero A, Vincenzini M (2003) Extracellular polysaccharide synthesis by Nostoc strains as affected by N source and light intensity. J Biotechnol 102(2):143–152

Parmar A et al (2011) Cyanobacteria and microalgae: a positive prospect for biofuels. Bioresour Technol 102:10163–10172

Pisciotta JM, Zou Y, Baskakov IV (2010) Light-dependent electrogenic activity of cyanobacteria. PLoS One 5(5):e10821

Saadatnia H, Riahi H (2009) Cyanobacteria from paddy fields in Iran as a biofertilizer in rice plants. Plant Soil Environ 55(5):207–212

Singh JS, Kumar A, Rai AN, Singh DP (2016) Cyanobacteria: a precious bio-resource in agriculture, ecosystem, and environmental sustainability. Frontiers in microbiology. 7:186282

Singh R, Parihar P, Singh M, Bajguz A, Kumar J, Singh S et al (2017) Uncovering potential applications of cyanobacteria and algal metabolites in biology, agriculture and medicine: current status and future prospects. Front Microbiol 8:212348

Sjöblom M, Matsakas L, Krige A, Rova U, Christakopoulos P (2017) Direct electricity generation from sweet sorghum stalks and anaerobic sludge. Ind Crop Prod 108:505–511

Song T, Mårtensson L, Eriksson T, Zheng W, Rasmussen U (2005) Biodiversity and seasonal variation of the cyanobacterial assemblage in a rice paddy field in Fujian, China. FEMS Microbiol Ecol 54(1):131–140

Sreeharsha RV, Mohan SV (2020) Obscure yet promising oleaginous yeasts for fuel and chemical production. Trends Biotechnol 38(8):873–887

Sreeharsha RV, Dubey N, Mohan SV (2023) Orienting biodiesel production towards sustainability and circularity by tailoring the feedstocks and processes. J Clean Prod:137526

Vargas SR et al (2018) Optimization of biomass and hydrogen production by Anabaena sp.(UTEX 1448) in nitrogen-deprived cultures. Biomass Bioenergy 111:70–76

Wang G, Jia Q, Li T, Dai S, Wu H, He H et al (2014) Bacterioplanes sanyensis gen. nov., sp. nov., a PHB-accumulating bacterium isolated from a pool of Spirulina platensis cultivation. Arch Microbiol 196:739–744

Wang J et al (2021) A review on polyhydroxyalkanoate production from agricultural waste biomass: development, advances, circular approach, and challenges. Bioresour Technol 342:126008

Wei J, Liang P, Huang X (2011) Recent progress in electrodes for microbial fuel cells. Bioresour Technol 102(20):9335–9344

Yodsang P et al (2018) Factors affecting photobiological hydrogen production in five filamentous cyanobacteria from Thailand. Photosynthetica 56:334–341

Part III
Avenues and Opportunities

Chapter 8
Genome Mining and Metabolic Engineering of Photosynthetic Microbes for Value Addition

Abstract Omics technologies including genomics, transcriptomics, proteomics, and metabolomics are essential tools to study the genes, metabolic pathways, and biochemical processes of photosynthetic microbes. Large-scale sequencing of genome and transcriptome has revolutionized microbial photosynthesis research, providing global snapshots of cellular behavior under different conditions. The current chapter highlights the principles and applications of genomics in microalgae and cyanobacteria research, and how these techniques contribute to developing genetically engineered microbes. The chapter also highlights the significance of systems-level technologies in elucidating the control and integration of metabolic pathways, offering potential targets for enhancing the synthesis of value-added chemicals. The advancement of novel and exceptionally effective phenotypes via the production of transgenic microalgae has been expedited by the deliberate targeting of genes employing reverse or forward genetic methodologies. Numerous advancements have been achieved, including enhanced lipid and carbohydrate synthesis, elevated H_2 generation, and the redirection of metabolic flow toward certain final outputs. The manipulation of lipid metabolism and the modification of sugar metabolism that is resistant to alteration have the potential to significantly increase both biomass output and lipid yields. The development of modern metabolic engineering tools in photosynthetic microbes has the potential to create cell factories with diverse applications.

Keywords Biofuel feedstock · Genomics · Genetic engineering · Metabolic engineering · Lipid metabolism · CRISPR-Cas

8.1 Introduction

In the preceding chapters, copious applications of photosynthetic microbes within the biorefinery framework have been examined. Henceforth, understanding their molecular responses to different environmental cues and strategies to engineer their metabolism is essential for achieving economic viability in the bioprocess.

R. V. Sreeharsha, S. Venkata Mohan, *Microbial Photosynthesis*,
https://doi.org/10.1007/978-981-97-1253-3_8

Identifying new photosynthetic microbes with novel functions and improving the viability of wild-type strains will address big questions like how to overcome environmental stress factors and growth-related issues. In photosynthetic microbes, the success of a bioprocess usually depends on three factors: (1) the photosynthetic efficiency of the green cell to accumulate carbon which subsequently determines the biomass production, (2) the tolerance of growth inhibitors and toxic elements in the culture media and utilization recalcitrant sugars to sustain productivity, and (3) distribution of carbon flux toward targeted end products to serve specific purpose of biorefinery (Sreeharsha and Mohan 2020). Hence, the newly progressing omics and genetic engineering approaches should focus on suitable genetic targets to achieve the specific output such as achieving the maximum photosynthetic efficiency or maximal solar-energy conversion or producing energy-efficient phenotypes.

Omics technologies such as genomics, transcriptomics, proteomics, and metabolomics help to study the gene and metabolic regulation of biochemical pathways. These system-level technologies can decode the metabolic flux that could help to improve the performance of wild-type strains for value-added compounds production through genetic engineering. Genomics focuses on the study of nucleotide sequences including DNA and complementary DNA (cDNA) and gives a basic understanding of the organization of the genome and its structural and functional status in an organism. Transcriptomics deals with understanding the functionality of a cell under various growth conditions by analyzing the expression and abundance of RNA transcripts. Unraveling the complete genome of photosynthetic microbes especially smaller genomes of unicellular microbes is rather simple when compared to the complex genomes of higher eukaryotes. Furthermore, the comprehensive study of gene function across different organisms provides insights into evolutionary characteristics of conserved genes and is crucial to phylogenetic studies.

Proteomics and metabolomics validate the transcriptome data by investigating the phenotypic alterations occurring inside cells as a result of genetic and environmental influences. Proteomics encompasses the investigation of protein expression patterns and the interactions between proteins, whereas metabolomics centers on the analysis of low molecular weight biomolecules, including carbohydrates, lipids, and secondary metabolites. In a way, they detect and quantify value-added compounds including carotenoids, steroids, lectins, polypeptides, polysaccharides, and toxins that have medicinal, nutritive, and industrial applications. Also, post-translational modifications, complex multiprotein structures, localization of polypeptide chains could be studied using proteomics techniques. Interactomics is an advanced approach amalgamating different omics and physiology techniques to understand the regulatory mechanisms of metabolic pathways. This strategy can significantly improve value-added compound production by regulating metabolic flux and interconversions between molecules (Rodríguez-Moyá and Gonzalez 2010; Schenk et al. 2008). Similarly, *Algomics* is the term introduced to study the systemic responses of microalgae in an integrated fashion.

Significant advances in next-generation sequencing technologies have revolutionized microbial photosynthesis research by facilitating large-scale sequencing of genome and transcriptome and rapid deposition of data in biological databases (Nag Dasgupta et al. 2018; Remmers et al. 2018; Salama et al. 2019a). Accessibility to genome information via public databases enables better understanding of the metabolic pathways and facilitate genetic manipulation for industrial applications (Guarnieri et al. 2011; Jamers et al. 2009). Omics has an important role in understanding structural and functional aspects of genomes and metabolic pathways, thus providing global snapshots of cellular behavior under different environmental conditions (Salama et al. 2019b; Blaby-Haas and Merchant 2019; Zhu et al. 2008; El-Sheekh et al. 2022; Radakovits et al. 2010; Sproles et al. 2021; Wijffels and Barbosa 2010). This chapter provides an in-depth analysis of the latest advancements in omics technologies applied to cyanobacteria and microalgae. The primary focus is on exploring potential genetic engineering strategies that can enhance these microorganisms' capabilities as a biorefinery platform for the synthesis of bioproducts. The approaches discussed in this chapter are valuable clues for modulating metabolic pathways of photosynthetic microbes, especially non-model microbes to produce value-added compounds and to utilize waste resources.

8.2 Genomics: Principles and Applications

Whole genome sequencing is largely helpful in functional characterization of newly isolated species and also enables comparative study of genomes and metabolic pathways across species. The genome of reference organisms has been particularly useful for mapping the growth-specific adaptations of non-model strains at genetic and molecular levels. High-throughput sequencing of whole genomes will identify novel synthesis pathways, target sites for mutation or transgenic studies, and syntenic gene patterns. High-quality genomes facilitate a better understanding of promoters and regulatory elements, intron/exon structure, centromeres, epigenomics, and a complete repertoire of genes. Of late, whole-genome and transcriptome sequences become invaluable resources for designing genetic and metabolic engineering experiments by circumventing the concerns about the complexity of genetic elements and tools (Blaby-Haas and Merchant 2019; Schneeberger and Weigel 2011; Chen et al. 2016; Kumar et al. 2017).

The basic requirement for a WGS or transcriptomics experiment is to isolate high-quality genomic DNA or RNA converted into cDNA, respectively (Sreeharsha et al. 2022). There are established protocols for genomics or cDNA library preparation for sequencing needs according to the manufacturer's instructions. The popular sequencing platforms for WGS and transcriptomics studies include Illumina-HiSeq, MiSeq, TrueSeq, and NovaSeq which vary based on the data output, read length, and efficiencies. The raw genome sequence in the form of scaffolds or reads gives little insight into biological function by itself, and it needs to be decoded by performing structural annotation. Several bioinformatics tools enable the assembly and annotation of short reads into expressed sequence data, particularly for non-model organisms without a known reference. The parameters such as mean length of unigenes, GC content, and N50 value indicate the coverage and depth of the sequencing, while FPKM and RPKM values indicate the quality of sequenced raw reads. Assembly and structural annotations specify the genomic features such as genes, coding and non-coding regions, transcription start and stop sites, and alternative splice sites. The sequencing also gives information on SSR markers, SNPs, repetitive DNA, trans-splicing, and post-transcriptional modifications (Sreeharsha et al. 2016). The completeness and accuracy of structural annotations vary across species depending on the complexity of genome and its size and quality of raw reads. In the post-genomic era, research efforts are focused on the development of tools to decode the functional significance of specific sequences in the context of biological mechanisms (Blaby-Haas and Merchant 2019). Whole genome sequencing can provide thousands of nuclear markers for phylogenetic and population-level studies enabling genome-wide investigations into fundamental aspects. *Metagenomics* is a powerful sequencing approach that involves the sequencing of target genes or complete genomes from heterogeneous mixed populations (Zeyaullah et al. 2009). This approach is particularly important in identifying individual bacterial species from a mixed population (Cissell and McCoy 2021; Sanseverino et al. 2022).

8.2.1 Current Scenario of Microalgae Genomics

The unicellular, water-oxidizing green alga *Chlamydomonas reinhardtii* was studied extensively for genomic resources and emerged as a model microalga. In particular, the genomic features of photosynthesis, flagellar function, phototaxis, nutrient acquisition, and the biosynthesis of lipids were explored in this organism through different approaches of omics techniques (Grossman et al. 2007). A draft genome followed by the whole genome sequence was made available in public databases, a decade ago and further deepened studies also revealed organellar genomes (Merchant et al. 2007). Other than *C. reinhardtii,* the eukaryotic photosynthetic microbes for which genomic resources are available through NGS are *Dunaliella* spp., *Chlorella* spp., *Volvox carteri* (green alga), *Cyanidioschyzon merolae* (red alga), *Ostreococcus lucimarinus*, *Ostreococcus tauri*, *Aureococcus anophagefferens* (a harmful algal bloom component), *Phaeodactylum tricornutum*, *Thalassiosira pseudonana* (diatoms), *Micromonas pusilla*, *Fragilariopsis cylindrus*, *Pseudo-nitzschia*, *Thalassiosira rotula*, *Botryococcus braunii*, *Micromonas pusilla*, *Galdieria sulphuraria*, and *Porphyra purpurea.* In addition, there are several ongoing efforts to sequence whole and organellar genomes, as well as dynamic transcriptomes for many different industrially important microalgae. The potential of *Scenedesmus quadricauda* LWG002611, a member of Chlorophyceae class, and order Sphaeropleales for biodiesel production has been revealed by whole genome sequencing (Nag Dasgupta et al. 2018).

8.2.2 Current Scenario of Cyanobacteria Genomics

The model cyanobacterial species, *Synechocystis* which is a freshwater, non-filamentous, non-nitrogen-fixing organism and capable of both photoautotrophic and heterotrophic growth was studied extensively for genetic resources. The genome of *Synechocystis* Sp. PCC 6803 was completely sequenced revealing crucial elements of genetic regulation and photosynthesis. The availability of genomic resources, coupled with physiological and biochemical data has made *Synechocystis* sp. PCC 6803 strain one of the most popular organisms for genetic and physiological studies of photosynthesis. Among Cyanobacteria, over 20 species are with complete genome sequences (http://genome.jgipsf.org/mic_cur1.html) and over 30 species are in progress.

8.3 Gene Editing and Engineering—Principles and Applications

Isolation of superior strains and their characterization through physiological and molecular approaches is the traditional approach in developing novel and industrial strains. This depends on their ability to natural adaptation to specific condition and to produce targeted product (Ranadheer et al. 2019; Sreeharsha et al. 2023). Despite the efforts in isolating novel and superior microbial species with higher efficiency of photosynthesis, the realistic productivities are not reaching the theoretical maximums due to the influence of growth and climatic conditions making the bioprocess uneconomical. To overcome this, there are intensive global research efforts aimed at increasing and modifying the accumulation of targeted products such as lipids, alcohols, hydrocarbons, polysaccharides, and other energy storage compounds in photosynthetic organisms through genetic engineering. The genomic data generated through omics technologies could be used subsequently to improve microbial strains through genetic engineering as a way to improve traits suitable for industrial applications.

Prior to the development of advanced genetic engineering techniques, the strain improvement has been done using UV mutagenesis, insertional or random mutagenesis, and targeted gene disruption which is considered as non-GMO approaches. Though they are not successful in obtaining targeted modifications, the resultant mutants won't have any regulatory problems of GMO. In such cases, the omics technologies play an important role in probing the useful mutants through sequencing and annotation of genomes and subsequent transcriptome analysis. Another successful approach in modulating the gene expression of targeted genes is transcription factor engineering wherein the overexpressed TFs can influence the expression of a set of genes involved in a metabolic pathway. Transcription factors or enhancers/activators are the DNA binding proteins which regulates the gene expression through DNA–protein interactions. This approach is particularly helpful in either enhancing or suppressing the gene expression without changing the gene sequence (Courchesne et al. 2009; Mochdia and Tamaki 2021). The transcriptional regulation of gene expression mediated by miRNA, siRNA, and other non-coding RNAs as well as RNAi techniques is proved to be the efficient approaches for functional genomics studies in *C. reinhardtii* and other non-model species (Wang et al. 2017).

The new age genome editing techniques involve the specific modification of the gene sequences so as to enhance or block certain metabolic pathways, thus changing the metabolic flux of the cell. In cyanobacteria and microalgae, several types of genomes editing systems have been developed until now, including zinc-finger nuclease (ZFN), transcription activator like endonucleases (TALENs), and CRISPR/Cas9 (Jeon et al. 2017; Jeong et al. 2023; Weyman et al. 2015; Zhang et al. 2019). Genome editing of marine diatoms *Phaeodactylum tricornutum* with TALEN as well as CRISPR/Cas9 and *Thalassiosira pseudonana* with CRISPR/Cas9, and marine eustigmatophyte *Nannochloropsis* spp. with CRISPR/Cas9 has been successfully demonstrated (Ajjawi et al. 2017; Daboussi et al. 2014; Hopes et al. 2016; Nymark et al. 2016; Serif et al. 2017; Wang et al. 2016). Successful generation of engineered strains depends on the optimized protocols for genetic transformation and subsequent selection. Compared with the large numbers of genetically transformed bacteria, fungi, and terrestrial plants, the number of algae that can be transformed is low. The first successful transformation of a microalgal strain was achieved in 1994, with *Cyclotella cryptica* and *Navicula saprophila* which showed improved biodiesel production (Dunahay et al. 1995). The technique utilized particle bombardment and an antibiotic resistance selectable marker under the control of the ACCase promoter and terminator elements. Previously, ACCase and UGPase genes were isolated and characterized, and further the expression levels were altered through transformation techniques. Evern after several decades of research, only few microalgal strains are benefited with optimized genetic transformation tools. *Chlamydomonas reinhardtii*, *Phaeodactylum tricornutum*, *Chlorella kessleri*, *Porphyridium Sp.*, and *Dunaliella salina* are among the few other microalgal strains which have been successfully transformed (Faktorová et al. 2020; Feng et al. 2014). Application of phenolic compounds such as cinnamic acid, coumarin, and vanillin in efficient agrobacterium-mediated transformation was demonstrated in *Nannochloropsis* (Cha et al. 2011). The following topics in the current chapter cover the genetic engineering strategies currently in use and potential targets for genome editing for various applications in photosynthetic microbes (Table 8.1). Furthermore, Fig. 8.1 presents a thorough perspective on several potential approaches of genetic engineering for strain improvement.

8.3.1 For Nutraceuticals

The pigments produced from photosynthetic microbes including β-carotene, lutein, astaxanthin, zeaxanthin are having high economic value as they can be used as nutraceuticals. Similar to the plant derived nutraceuticals, microalgae-derived pigments also become popular due to their natural origin. The flux in the carotenoids biosynthetic pathway is tightly regulated toward specific pigment depending on the environmental condition (Refer Chap. 1). The expression of corresponding enzymes is crucial to upregulate a specific compound as all the carotenoids originated from the same starting material. Certain strains of microalgae and cyanobacteria are

Table 8.1 Summary of recent genome editing in photosynthetic microbes and the targeted pathways

Organism	Targeted pathway	Editing strategy	References
C. reinhardtii CC125	Lipid metabolism	Overexpression of acetyl-CoA synthetase	Rengel et al. (2018)
C. reinhardtii CC400	Lipid metabolism	CRISPRi-based silencing of PEP-carboxylase	Kao and Ng (2017)
C. reinhardtii		CRISPR-Cas9	Zhang et al. (2020)
N. oceanica	Lipid accumulation	CRISPR-Cas9	Hu et al. (2021)
			Li et al. (2020)
Phaeodactylum tricornutum (diatom microalgae)	Lipid biosynthesis	CRISPR-Cas9	Nymark et al. (2017)
	Carbon fixation		Stukenberg et al. (2018)
Haematococcus pluvialis	Astaxanthin	CRISPR-Cas9	Kang et al. (2020)
			Xu et al. (2021)
Dunaliella salina (halotolerant microalgae)	Beta carotene, a precursor of Vitamin A	CRISPR-Cas9	Hu et al. (2021)
Synechococcus elongatus	Biofuels and other valuable compounds	CRISPR-Cas9	Gao et al. (2016)
			Li et al. (2018)
Tetraselmis suecica	Lipid biosynthesis and biomass productivity	CRISPR-Cas9	Wang et al. (2019)
Euglena gracilis	Biofuels and other valuable compounds	CRISPR-Cas9	Yamada et al. (2020)
Scenedesmus obliquus	Triacylglycerols increase	CRISPR-Cas9	Wei et al. (2019)
Chlorococcum sp	Lipid biosynthesis and stress response	CRISPR-Cas9	Wang et al. (2019)
Chlorella vulgaris	Lipids and pigments for cosmetics	CRISPR-Cas9	Yin et al. (2020)
Porphyridium purpureum	Phycobiliproteins	CRISPR-Cas9	Jeon et al. (2021)
Dunaliella tertiolecta (halotolerant microalgae)	Salt tolerance and carotenoid biosynthesis	CRISPR-Cas9	Wang et al. (2019)
Phaeodactylum tricornutum		TALENs	Nymark et al. (2016)
Nannochloropsis oceanica		CRISPR-Cpf1 (Cas12a)	Li et al. (2018)
Nannochloropsis oceanica		Oligonucleotide directed mutagenesis	Kilian et al. (2011)

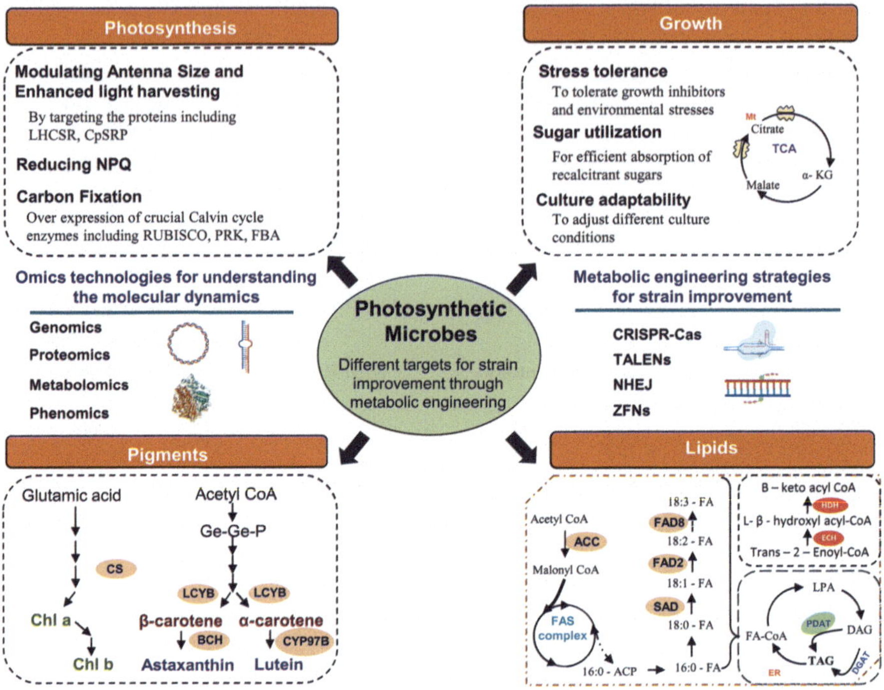

Fig. 8.1 Genome editing strategies and important targets for editing in photosynthetic microbes

known to produce specific pigments, and the strains can be engineered further to enhance the pigment accumulation. Lycopene beta cyclase and carotene beta hydroxylase are the prominent rate limiting enzymes which can be upregulated to enhance lutein and zeaxanthin production, respectively. In addition, downregulation of β-carotene cyclase upregulates the downstream product formation.

8.3.2 *For Growth and Biomass Enhancement*

Photosynthesis plays a significant role in growth and biomass accumulation of microalgae and cyanobacteria. The light and dark reactions concomitantly operate to accumulate the organic sugars which ultimately support the growth of the organism. Of late, several studies observed the enhancement in photosynthesis by modulating the expression of extrinsic proteins of photosystems so as to enhance the light capture, electron transport chain, and photochemical quenching. Light harvesting complexes are the dynamic sources of engineering that significantly influence the growth of the organism. Furthermore, engineering recalcitrant sugar metabolism will enhance the sugar uptake potential of microbes and enhance the biomass productivity especially in heterotrophic conditions. This has a potential value in valorizing the waste resources which contain non-conventional sugars such as xylose and arabinose which are not suitable for microbial uptake. Also, microbes which are resistant to growth inhibitors including hydroxymethylfurfural (HMF) and furfural that are released during the hydrolysis of waste resources will have great value in waste valorization. There are several studies reporting the engineering of microalgae which are robust to growth inhibitors and can be able to feed on recalcitrant sugars through structural modifications of cell wall composition and by expressing enzymes that can metabolize the targeted sugars.

8.3.3 For Biofuels

Microalgae accumulate triacylglycerols with unsaturated fatty acids which are ideal sources of biodiesel production through the transesterification process. Globally, there has been increasing attention to identify potential microalgal strains that can produce superior lipid yields by simultaneously utilizing wastewaters. Engineering lipid metabolism along with other high-value products has the potential to design an industrial strain that is useful for biorefinery applications. Salt and temperature tolerance, resistance to external environmental stresses, ability to grow at high cell density, and ease of harvesting are the other ideal characteristics to make a robust microbial strain (Larkum et al. 2012). Simultaneously, robust lipid producing strains with high biomass productivity can be generated by targeting photosynthetic pathway. A comprehensive knowledge of precursors, intermediates, and end products of lipid metabolism, and the identification of their regulatory networks that control metabolic flux are crucial to design specific engineering strategy for optimized biodiesel production. Attempts have already been made to enhance lipid productivity and to modulate the fatty acid composition in microalgae by understanding the regulation of lipid biosynthetic pathway from higher plants (Sreeharsha et al. 2023; Hu et al. 2008). Though the photosynthetic microbes possess many homologs of lipid biosynthetic genes from terrestrial plants, their dynamic molecular regulation and structural features were not studied thoroughly. Unraveling the extensive genomic and transcriptomic data of diverse photosynthetic microbes will be useful in this regard (Rismani-Yazdi et al. 2011; Salama et al. 2019a). The data will provide valuable cues in understanding the dynamics of lipid metabolism which will ultimately facilitate improvement of biodiesel production. It is presumed that the strategies that are successful in improving the quality and composition of lipids in higher plants will also work well in photosynthetic microbes especially in microalgae. Furthermore, the advent of CRISPR-mediated genetic engineering facilitates introduction of regulatory genetic elements of lipid metabolism from higher plants into microbes to enhance the kinetics of the lipid metabolic enzymes (Radakovits et al. 2010; Sreeharsha and Mohan 2020). While the direct approaches to enhance lipid quantity and composition of fatty acids in microbes include engineering fatty acid chain length and truncating the β-oxidation process (Blatti et al. 2013). However, it is to be noted that the engineered strains should retain high photosynthetic rates, better resource utilization, and high biomass productivity in order to create strains capable of producing high-quality biodiesel at an industrial scale (Sreeharsha et al. 2023).

Chlamydomonas reinhardtii followed by *Chlorella vulgaris* was considered to be the most studied microalgae among the few oleaginous microalgae that are benefited from genetic and metabolic engineering approaches (Chiu et al. 2009; Grobbelaar 2010; Oh et al. 2009; Rodolfi et al. 2009). Their amenability as model organisms for genetic engineering experiments is due to the availability of the successful transformation protocols. In addition to biofuel and bioenergy applications, there are reports that show the potential of these microalgae for hydrogen production. In addition to these microalgal strains, marine diatoms are also reported to be potential strains for modulating lipid metabolism toward superior biodiesel production. Improvement of lipid production in *Nannochloropsis oculata,* a eustigmatophyte alga of Haptophyte group, was achieved through biochemical manipulation nitrogen starvation and CO_2 enrichment, thus enhancing the natural ability of producing the storage lipid globules (Chiu et al. 2009). Knocking out single transcription regulator gene (a homolog of fungal Zn(ii)2 Cys6-encoding genes) through genome editing, doubled the lipid productivity in marine microalga *N. gaditana* (Ajjawi et al. 2017; Maeda et al. 2018). The clues obtained from these model photosynthetic microbial strains could help to transform other species in the near future.

8.4 Carboxysome Engineering in Cyanobacteria

Bacterial microcompartments are a type of structural and physiological optimization process to execute spatial and temporal regulation of metabolic reactions. Carboxysomes are one of such specialized microcompartment structures in cyanobacteria which are vital for cell survival in the natural environment. Carboxysomes which usually exist as multiprotein supercomplexes encapsulate crucial photosynthetic enzymes such as carbonic anhydrase and RUBISCO, thus improving the CO_2 fixation in cells (Fig. 8.2). They also facilitate CO_2 concentrating mechanisms by coordinating inorganic carbon (Ci) transporters in plasma membranes and CO_2 uptake complexes in thylakoid membranes. Removing or impairing carboxysomes could render cells incapable of CO_2 fixation in ambient air and thus lead to cell death. Carboxysomes occurring in cyanobacteria can be categorized into two types α-carboxysomes and β-carboxysomes which found in oceanic and freshwater cyanobacterial strains, respectively (Rae et al. 2013). Carboxysomes show excellent self-assembly and modularity due to which they are very much amenable for synthetic biology approaches. Synthesizing functional carboxysomes with a streamlined composition are highly advantageous for industrial applications. The structural features of the Rubisco and other proteins need to be optimized while designing the composition of the carboxysomes.

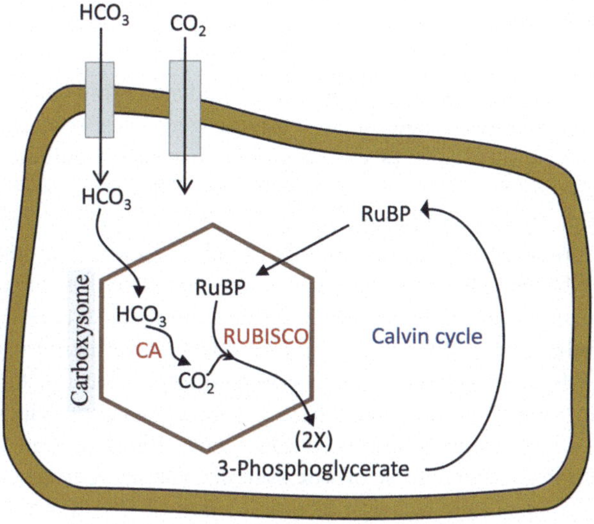

Fig. 8.2 Schematic representation of the carboxysomes and their structural aspects in cyanobacteria

Carboxysome operon from *H. neapolitanus* was expressed to produce recombinant carboxysome-like structures with CO_2 fixation activity in *Escherichia coli* (Bonacci et al. 2012). Furthermore, key carboxysome genes including ccmK1, ccmK2, ccmL, and ccmO from *Halothece* sp. PCC 7418's expressed in *E. coli* to produce synthetic α-carboxysome shells (Cai et al. 2016). The operon has also been utilized to recreate carboxysomes in *Corynebacterium glutamicum*, a gram-positive bacterium (Baumgart et al. 2017). Due to the numerous operons that code for β-carboxysome proteins, reconstructing them in prokaryotic species lags considerably behind α-carboxysome engineering. Recently, it was demonstrated that all of the carboxysome proteins could be expressed in E. coli and form carboxysome-like structures with detectable CO_2-fixing activity (Fang et al. 2018). This is a step toward building functional β-carboxysomes in eukaryotic organisms, including plants. The potential for creating chimeric carboxysomes is increased by the common structural characteristics and high sequence similarity of carboxysome proteins from many hosts. It has been shown that the form 1A Rubisco large subunit CbbL and shell protein CsoS1A of carboxysomes from *H. neapolitanus* could be individually incorporated into the natural carboxysomes in Syn7942, proving the interchangeability of carboxysome building components (Fang et al. 2018; Cai et al. 2016). Expressing carboxysomes into plants to boost photosynthesis and crop yields has generated a great deal of interest due to its capacity to enhance carbon assimilation (Hanson et al. 2016). The model crop *Nicotiana benthamiana* has had its chloroplasts modified to express carboxysome components, such as transiently expressing carboxysome shell proteins and Syn7942 Rubisco with a rapid carboxylation rate to substitute native Rubisco (Lin et al. 2014; Occhialini et al. 2016).

8.5 Conclusions

Omics technologies play an important role in understanding photosynthesis and other metabolic pathways in microbes. Implying genomics, transcriptomics, proteomics, and metabolomics can provide a global snapshot of cellular behavior and help to identify target genes for genetic engineering. They are also effective in optimizing microbial growth and production under different abiotic stress conditions such as nutrient repletion/depletion, phytohormones, salinity, heat stress, light, UV radiation, etc. Microalgae and cyanobacteria are considered as amenable genetic models for manipulation of metabolic pathways through advanced genetic engineering approaches for enhancing targeted products and utilizing waste resources using reverse or forward genetic strategies. In addition to direct metabolic pathway modification, metabolic engineering can also target the manipulation of individual enzyme kinetics, molecular transport, product secretion, cell wall structure, and regulatory functions within photosynthetic systems. Whole-genome sequencing and comparative genomics, together with the application of CRISPR-Cas systems and other genome-engineering technologies, will enable photosynthetic microbes to ascend to the level of industrial workhorses. However, there is a need for optimization of transformation protocols for several non-model organisms to effectively undertake genetic manipulations.

References

Ajjawi I, Verruto J, Aqui M, Soriaga LB, Coppersmith J, Kwok K, Peach L, Orchard E, Kalb R, Xu W (2017) Lipid production in *Nannochloropsis gaditana* is doubled by decreasing expression of a single transcriptional regulator. Nat Biotechnol 35(7):647–652

Baumgart M, Huber I, Abdollahzadeh I, Gensch T, Frunzke J (2017) Heterologous expression of the Halothiobacillus neapolitanus carboxysomal gene cluster in Corynebacterium glutamicum. J Biotechnol 258:126–135

Blaby-Haas CE, Merchant SS (2019) Comparative and functional algal genomics. Annu Rev Plant Biol 70:605–638

Blatti JL, Michaud J, Burkart MD (2013) Engineering fatty acid biosynthesis in microalgae for sustainable biodiesel. Curr Opin Chem Biol 17(3):496–505

Bonacci W, Teng PK, Afonso B, Niederholtmeyer H, Grob P, Silver PA, Savage DF (2012) Modularity of a carbon-fixing protein organelle. Proc Natl Acad Sci 109(2):478–483

Cai F, Bernstein SL, Wilson SC, Kerfeld CA (2016) Production and characterization of synthetic carboxysome shells with incorporated luminal proteins. Plant Physiol 170(3):1868–1877

Cha T-S, Chen C-F, Yee W, Aziz A, Loh S-H (2011) Cinnamic acid, coumarin and vanillin: alternative phenolic compounds for efficient agrobacterium-mediated transformation of the unicellular green alga, *Nannochloropsis sp.* J Microbiol Methods 84(3):430–434

Chen W, Wang W, Peng M, Gong L, Gao Y, Wan J, Wang S, Shi L, Zhou B, Li Z (2016) Comparative and parallel genome-wide association studies for metabolic and agronomic traits in cereals. Nat Commun 7(1):12767

Chiu S-Y, Kao C-Y, Tsai M-T, Ong S-C, Chen C-H, Lin C-S (2009) Lipid accumulation and CO_2 utilization of Nannochloropsis oculata in response to CO_2 aeration. Bioresour Technol 100(2):833–838

Cissell EC, McCoy SJ (2021) Shotgun metagenomic sequencing reveals the full taxonomic, trophic, and functional diversity of a coral reef benthic cyanobacterial mat from Bonaire, Caribbean Netherlands. Sci Total Environ 755:142719

Courchesne NMD, Parisien A, Wang B, Lan CQ (2009) Enhancement of lipid production using biochemical, genetic and transcription factor engineering approaches. J Biotechnol 141(1–2):31–41

Daboussi F, Leduc S, Maréchal A, Dubois G, Guyot V, Perez-Michaut C, Amato A, Falciatore A, Juillerat A, Beurdeley M (2014) Genome engineering empowers the diatom Phaeodactylum tricornutum for biotechnology. Nat Commun 5(1):3831

Dunahay TG, Jarvis EE, Roessler PG (1995) Genetic transformation of the diatoms cyclotella cryptica and navicula saprophila. J Phycol 31(6):1004–1012

El-Sheekh M, El-Dalatony M, Thakur N, Zheng Y, Salama E-S (2022) Role of microalgae and cyanobacteria in wastewater treatment: genetic engineering and omics approaches. Int J Environ Sci Technol 19:2173–2194

Faktorová D, Nisbet RER, Fernández Robledo JA, Casacuberta E, Sudek L, Allen AE, Ares M Jr, Aresté C, Balestreri C, Barbrook AC (2020) Genetic tool development in marine protists: emerging model organisms for experimental cell biology. Nat Methods 17(5):481–494

Fang Y, Huang F, Faulkner M, Dykes GF, Liu LN (2018) Engineering and modulating functional cyanobacterial CO2-fixing organelles. Front Plant Sci 9:356777

Feng S, Li X, Xu Z, Qi J (2014) Dunaliella salina as a novel host for the production of recombinant proteins. Appl Microbiol Biotechnol 98:4293–4300

Gao X, Sun T, Pei G, Chen L, Zhang W (2016) Cyanobacterial chassis engineering for enhancing production of biofuels and chemicals. Appl Microbiol Biotechnol 100(8):3401–3413

Grobbelaar JU (2010) Microalgal biomass production: challenges and realities. Photosynth Res 106:135–144

Grossman AR, Croft M, Gladyshev VN, Merchant SS, Posewitz MC, Prochnik S, Spalding MH (2007) Novel metabolism in chlamydomonas through the lens of genomics. Curr Opin Plant Biol 10(2):190–198

Guarnieri MT, Nag A, Smolinski SL, Darzins A, Seibert M, Pienkos PT (2011) Examination of triacylglycerol biosynthetic pathways via de novo transcriptomic and proteomic analyses in an unsequenced microalga. PLoS One 6(10):e25851

Hanson DT, Stutz SS, Boyer JS (2016) Why small fluxes matter: the case and approaches for improving measurements of photosynthesis and (photo) respiration. J Exp Bot 67(10):3027–3039

Harada R, Nomura T, Yamada K, Mochida K, Suzuki K (2020) Genetic engineering strategies for Euglena gracilis and its industrial contribution to sustainable development goals: A review. Front bioeng biotechnol 14(8):790

Hopes A, Nekrasov V, Kamoun S, Mock T (2016) Editing of the urease gene by CRISPR-Cas in the diatom *Thalassiosira pseudonana*. Plant Methods 12(1):1–12

Hu L, Feng S, Liang G, Du J, Li A, Niu C (2021) CRISPR/Cas9-induced β-carotene hydroxylase mutation in Dunaliella salina CCAP19/18. AMB Express. 11(1):83

Hu Q, Sommerfeld M, Jarvis E, Ghirardi M, Posewitz M, Seibert M, Darzins A (2008) Microalgal triacylglycerols as feedstocks for biofuel production: perspectives and advances. Plant J 54(4):621–639

Jamers A, Blust R, De Coen W (2009) Omics in algae: paving the way for a systems biological understanding of algal stress phenomena? Aquat Toxicol 92(3):114–121

Jeon MS, Han SI, Jeon M, Choi YE (2021) Enhancement of phycoerythrin productivity in Porphyridium purpureum using the clustered regularly interspaced short palindromic repeats/CRISPR-associated protein 9 ribonucleoprotein system. Bioresource technology. 330:124974

Jeon S, Lim J-M, Lee H-G, Shin S-E, Kang NK, Park Y-I, Oh H-M, Jeong W-J, Jeong, B.-r., Chang, Y.K. (2017) Current status and perspectives of genome editing technology for microalgae. Biotechnol Biofuels 10(1):1–18

Jeong B-R, Jang J, Jin E (2023) Genome engineering via gene editing technologies in microalgae. Bioresour Technol 373:128701

Kang H, Lee Y, Bae M, Park YK, Lee JY (2020) Astaxanthin inhibits alcohol-induced inflammation and oxidative stress in macrophages in a sirtuin 1-dependent manner. J Nutr Biochem 85:108477

Kao PH, Ng IS (2017) CRISPRi mediated phosphoenolpyruvate carboxylase regulation to enhance the production of lipid in Chlamydomonas reinhardtii. Bioresour Technol 245:1527–1537

Kilian O, Benemann CS, Niyogi KK, Vick B (2011) High-efficiency homologous recombination in the oil-producing alga Nannochloropsis sp. Proceedings of the National Academy of Sciences. 108(52):21265–21269

Kumar S, Sreeharsha RV, Mudalkar S, Sarashetti PM, Reddy AR (2017) Molecular insights into photosynthesis and carbohydrate metabolism in Jatropha curcas grown under elevated CO_2 using transcriptome sequencing and assembly. Sci Rep 7(1):11066

Larkum AW, Ross IL, Kruse O, Hankamer B (2012) Selection, breeding and engineering of microalgae for bioenergy and biofuel production. Trends Biotechnol 30(4):198–205

Li B, Zeng C, Li W, Zhang X, Luo X, Zhao W et al (2018) Synthetic oligonucleotides inhibit CRISPR-Cpf1-mediated genome editing. Cell Rep 25(12):3262–3272

Lin MT, Occhialini A, Andralojc PJ, Parry MA, Hanson MR (2014) A faster Rubisco with potential to increase photosynthesis in crops. Nature 513(7519):547–550

Maeda Y, Yoshino T, Matsunaga T, Matsumoto M, Tanaka T (2018) Marine microalgae for production of biofuels and chemicals. Curr Opin Biotechnol 50:111–120

Merchant SS, Prochnik SE, Vallon O, Harris EH, Karpowicz SJ, Witman GB, Terry A, Salamov A, Fritz-Laylin LK, Marĕchal-Drouard L (2007) The Chlamydomonas genome reveals the evolution of key animal and plant functions. Science 318(5848):245–250

Mochdia K, Tamaki S (2021) Transcription factor-based genetic engineering in microalgae. Plan Theory 10(8):1602

Nag Dasgupta C, Nayaka S, Toppo K, Singh AK, Deshpande U, Mohapatra A (2018) Draft genome sequence and detailed characterization of biofuel production by oleaginous microalga *Scenedesmus quadricauda* LWG002611. Biotechnol Biofuels 11:1–15

Nymark M, Sharma AK, Sparstad T, Bones AM, Winge P (2016) A CRISPR/Cas9 system adapted for gene editing in marine algae. Sci Rep 6(1):24951

Nymark M, Sharma AK, Hafskjold MC, Sparstad T, Bones AM, Winge P (2017) CRISPR/Cas9 gene editing in the marine diatom Phaeodactylum tricornutum. Bio-protocol. 7(15):e2442

Occhialini A, Lin MT, Andralojc PJ, Hanson MR, Parry MA (2016) Transgenic tobacco plants with improved cyanobacterial Rubisco expression but no extra assembly factors grow at near wild-type rates if provided with elevated CO 2. Plant J 85(1):148–160

Oh SH, Han JG, Kim Y, Ha JH, Kim SS, Jeong MH, Jeong HS, Kim NY, Cho JS, Yoon WB (2009) Lipid production in Porphyridium cruentum grown under different culture conditions. J Biosci Bioeng 108(5):429–434

Radakovits R, Jinkerson RE, Darzins A, Posewitz MC (2010) Genetic engineering of algae for enhanced biofuel production. Eukaryot Cell 9(4):486–501

Rae BD, Long BM, Whitehead LF, Förster B, Badger MR, Price GD (2013) Cyanobacterial carboxysomes: microcompartments that facilitate CO2 fixation. J Mol Microbiol Biotechnol 23(4-5):300–307

Ranadheer P, Kona R, Sreeharsha RV, Mohan SV (2019) Non-lethal nitrate supplementation enhances photosystem II efficiency in mixotrophic microalgae towards the synthesis of proteins and lipids. Bioresour Technol 283:373–377

Remmers IM, Wijffels RH, Barbosa MJ, Lamers PP (2018) Can we approach theoretical lipid yields in microalgae? Trends Biotechnol 36(3):265–276

Rengel R, Smith RT, Haslam RP, Sayanova O, Vila M, León R (2018) Overexpression of acetyl-CoA synthetase (ACS) enhances the biosynthesis of neutral lipids and starch in the green microalga Chlamydomonas reinhardtii. Algal Res 31:183–193

Rismani-Yazdi H, Haznedaroglu BZ, Bibby K, Peccia J (2011) Transcriptome sequencing and annotation of the microalgae Dunaliella tertiolecta: pathway description and gene discovery for production of next-generation biofuels. BMC Genomics 12:1–17

Rodolfi L, Chini Zittelli G, Bassi N, Padovani G, Biondi N, Bonini G, Tredici MR (2009) Microalgae for oil: strain selection, induction of lipid synthesis and outdoor mass cultivation in a low-cost photobioreactor. Biotechnol Bioeng 102(1):100–112

Rodríguez-Moyá M, Gonzalez R (2010) Systems biology approaches for the microbial production of biofuels. Biofuels 1(2):291–310

Salama E-S, Govindwar SP, Khandare RV, Roh H-S, Jeon B-H, Li X (2019a) Can omics approaches improve microalgal biofuels under abiotic stress? Trends Plant Sci 24(7):611–624

Salama E-S, Roh H-S, Dev S, Khan MA, Abou-Shanab RA, Chang SW, Jeon B-H (2019b) Algae as a green technology for heavy metals removal from various wastewater. World J Microbiol Biotechnol 35:1–19

Sanseverino I, Pretto P, António DC, Lahm A, Facca C, Loos R, Skejo H, Beghi A, Pandolfi F, Genoni P (2022) Metagenomics analysis to investigate the microbial communities and their functional profile during cyanobacterial blooms in Lake Varese. Microb Ecol 83:850–868

Schenk PM, Thomas-Hall SR, Stephens E, Marx UC, Mussgnug JH, Posten C, Kruse O, Hankamer B (2008) Second generation biofuels: high-efficiency microalgae for biodiesel production. Bioenergy Res 1:20–43

Schneeberger K, Weigel D (2011) Fast-forward genetics enabled by new sequencing technologies. Trends Plant Sci 16(5):282–288

Serif M, Lepetit B, Weißert K, Kroth PG, Bartulos CR (2017) A fast and reliable strategy to generate TALEN-mediated gene knockouts in the diatom *Phaeodactylum tricornutum*. Algal Res 23:186–195

Sproles AE, Fields FJ, Smalley TN, Le CH, Badary A, Mayfield SP (2021) Recent advancements in the genetic engineering of microalgae. Algal Res 53:102158

Sreeharsha RV, Dubey N, Mohan SV (2023) Orienting biodiesel production towards sustainability and circularity by tailoring the feedstocks and processes. J Clean Prod 414:137526

Sreeharsha RV, Mohan SV (2020) Obscure yet promising oleaginous yeasts for fuel and chemical production. Trends Biotechnol 38(8):873–887

Sreeharsha RV, Mudalkar S, Reddy AR (2022) Genome sequencing and analysis uncover the regulatory elements involved in the development and oil biosynthesis of Pongamia pinnata (L.)—a potential biodiesel feedstock. Front Plant Sci 13:747783

Sreeharsha RV, Mudalkar S, Singha KT, Reddy AR (2016) Unravelling molecular mechanisms from floral initiation to lipid biosynthesis in a promising biofuel tree species, Pongamia pinnata using transcriptome analysis. Sci Rep 6:34315

Stukenberg D, Zauner S, Dell'Aquila G, Maier UG (2018) Optimizing CRISPR/Cas9 for the diatom Phaeodactylum tricornutum. Frontiers in plant science. 9:740

Wang C, Chen X, Li H, Wang J, Hu Z (2017) Artificial miRNA inhibition of phosphoenolpyruvate carboxylase increases fatty acid production in a green microalga Chlamydomonas reinhardtii. Biotechnol Biofuels 10(1):1–11

Wang Q, Lu Y, Xin Y, Wei L, Huang S, Xu J (2016) Genome editing of model oleaginous microalgae *Nannochloropsis* spp. by CRISPR/Cas9. Plant J 88(6):1071–1081

Wang XW, Hu LF, Hao J, Liao LQ, Chiu YT, Shi M, Wang Y (2019) A microRNA-inducible CRISPR–Cas9 platform serves as a microRNA sensor and cell-type-specific genome regulation tool. Nat Cell Biol 21(4):522–530

Wei L, Lee D, Law CT, Zhang MS, Shen J, Chin DWC et al (2019) Genome-wide CRISPR/Cas9 library screening identified PHGDH as a critical driver for Sorafenib resistance in HCC. Nat Commun 10(1):4681

Weyman PD, Beeri K, Lefebvre SC, Rivera J, McCarthy JK, Heuberger AL, Peers G, Allen AE, Dupont CL (2015) Inactivation of *Phaeodactylum tricornutum* urease gene using transcription activator-like effector nuclease-based targeted mutagenesis. Plant Biotechnol J 13(4):460–470

Wijffels RH, Barbosa MJ (2010) An outlook on microalgal biofuels. Science 329(5993):796–799

Xu X, Liu C, Wang Y, Koivisto O, Zhou J, Shu Y, Zhang H (2021) Nanotechnology-based delivery of CRISPR/Cas9 for cancer treatment. Adv Drug Deliv Rev 176:113891

Yin H, Yuan X, Luo L, Lu Y, Qin B, Zhang J et al (2020) Appropriate delivery of the CRISPR/Cas9 system through the nonlysosomal route: application for therapeutic gene editing. Adv Sci 7(14):1903381

Zeyaullah M, Kamli MR, Islam B, Atif M, Benkhayal FA, Nehal M, Rizvi M, Ali A (2009) Metagenomics—an advanced approach for noncultivable micro-organisms. Biotechnol Mol Biol Rev 4(3):49–54

Zhang Y-T, Jiang J-Y, Shi T-Q, Sun X-M, Zhao Q-Y, Huang H, Ren L-J (2019) Application of the CRISPR/Cas system for genome editing in microalgae. Appl Microbiol Biotechnol 103:3239–3248

Zhang L, Zhang H, Liu Y, Zhou J, Shen W, Liu L et al (2020) A CRISPR–Cas9 system for multiple genome editing and pathway assembly in Candida tropicalis. Biotechnol Bioeng 117(2):531–542

Zhu X-G, Long SP, Ort DR (2008) What is the maximum efficiency with which photosynthesis can convert solar energy into biomass? Curr Opin Biotechnol 19(2):153–159

Chapter 9
Industrial Symbiosis of Photosynthetic Microbes with Other Biorefineries to Achieve Circular Economy

Abstract Industrialization has led to increased competition between agricultural and industrial needs in urban areas, resulting in municipal solid waste accumulating in landfills. To address this issue, circular bioeconomy, urban metabolism, and biorefineries can redirect non-productive waste toward societal benefits. This approach aligns with the principles of a circular bioeconomy, which emphasizes efficient food production and minimizes waste generation. An interdisciplinary approach that harnesses insights from plant, animal, and microbial research, combined with artificial intelligence and bioprocessing, can create a sustainable, self-contained ecosystem. Industrial symbiosis, a collaborative approach that fosters the sharing of resources, materials, energy, and by-products among different industries, can create mutual economic, environmental, and social benefits. The global population surge is causing severe impacts on climate change, agricultural land utilization, and waste management, affecting food security. Nutrient cycles, such as nitrogen assimilation and phosphate recycling, are crucial for maintaining nutrient cycles and sustaining ecosystem balance. The dynamic interplay of microbial interactions is crucial in maintaining nutrient cycles and sustaining ecosystems.

Keywords Industrial symbiosis · Microalgae · Cyanobacteria · Urban cycles · Circular economy · Biorefinery

9.1 Introduction

The rise of industrialization has reconfigured land use in urban areas, intensifying the competition between agricultural and industrial requirements (Bamsey et al. 2009; Chunxiao and Hong 2008). Additionally, industrialization contributes to vast amounts of municipal solid waste accumulating in landfills, rendering the land unproductive and emitting greenhouse gases (Mohan et al. 2016). Constructing resilient, self-sufficient cities in the face of changing climatic dynamics has become a priority in the recent past. Circular bioeconomy, urban metabolism, and biorefineries redirect non-productive waste toward societal benefits, fostering eco-friendly

R. V. Sreeharsha, S. Venkata Mohan, *Microbial Photosynthesis*,
https://doi.org/10.1007/978-981-97-1253-3_9

urban infrastructure (Mohan et al. 2020). Urban metabolism (UM) examines resource consumption and flow direction, offering insights into the net energy and material demands of urban populations (Alberti 2008; Dijst et al. 2018; Gandy 2004; Restrepo and Morales-Pinzón 2018). It accounts for waste emissions, treatment, and recycling, illuminating the technical and socio-economic facets of urban life. Integrated waste management through waste biorefineries facilitates zero-waste policies and responsible consumption, aligning with the principles of the circular bioeconomy, which emphasizes efficient food production while minimizing waste generation through optimized use of renewable resources (Mohan et al. 2019). However, amidst unprecedented global pandemics, the standalone mature technologies may not be sufficient for essential needs of the society. The multifaceted challenges of urban areas necessitate a multidimensional, integrated solution. Intriguingly, an interdisciplinary approach that harnesses insights from plant, animal, and microbial research, combined with artificial intelligence and bioprocessing, can create a sustainable, self-contained ecosystem.

Photosynthesis by microalgae or cyanobacteria harnesses sunlight, carbon dioxide, and nutrients into biomass and valuable compounds. With the advent of waste utilization, resource valorization, and nutrient recovery, the carbon sequestration by microorganisms could be integrated with various biomass conversion processes in a biorefinery approach to produce a range of products, including biofuels, biochemicals, and materials. The concept aligns with the principles of a circular economy, as it aims to maximize the value extracted from biomass resources while minimizing waste generation. The biorefinery can become more advantageous and economical by maintaining industrial symbiosis in a collaborative approach. Industrial symbiosis is the new, emerging, and innovative approach that fosters the sharing of resources, materials, energy, and by-products among different industries, with the goal of creating mutual economic, environmental, and social benefits. This helps in turning waste from one bioprocess into valuable input for another, promoting the efficient use of resources, reducing environmental impact, and enhancing overall sustainability. Industrial symbiosis is often considered a key strategy in the pursuit of a circular economy, where resources are continuously looped through different production cycles, minimizing waste, and maximizing value. Industries can reduce waste disposal costs, access cheaper inputs, and improve resource efficiency. At the same time, environmental impacts are reduced through decreased waste generation and a lower need for virgin resources. By reusing resources that would otherwise be discarded, it contributes to environmental preservation by reducing pollution and conserving natural resources. Additionally, economic gains arise from cost savings, enhanced competitiveness, and the creation of new revenue streams. It can be achieved by innovative and sustainable practices to use waste streams and by-products.

Integration of different industries in symbiotic manner is quite successful in European countries and south Asian countries. For instance, Kalundborg Eco-Industrial Park, Denmark is one of the earliest and most well-known examples of industrial symbiosis. This park involves multiple companies sharing resources and waste products. A power plant supplies steam to neighboring industries, a pharmaceutical company's waste heat is used for fish farming, and gypsum from a power plant is used in cement production, among other collaborations. Othe successful venture is Symbiosis Center, Denmark that facilitates the exchange of excess heat, water, and raw materials between various industries, leading to reduced waste generation and more efficient resource use. Riihimäki Eco-Industrial Park, Finland collaborates on various fronts, including the utilization of waste heat, sharing of materials, and joint waste treatment. The park aims to enhance the circular economy principles and decrease the environmental footprint of its member companies. VinylPlus® is a voluntary commitment by the European PVC industry to create a sustainable development model for the PVC value chain. It involves recycling PVC waste to produce new products, effectively closing the loop and minimizing the impact of PVC production on the environment. There are several advantages in maintaining industrial symbiosis: (1) Resource efficiency and reduced waste generation, (2) cost savings for industries through shared resources and reduced disposal costs, (3) environmental benefits, including reduced emissions and conservation of resources, (4) creation of new business opportunities and revenue streams, and (5) enhanced competitiveness and reputation for participating industries. Integration of different industries and technologies requires careful planning and coordination. Different industries may have varying production cycles and resource needs, thus identifying suitable matches for resource exchange becoming complex. Also, legal, regulatory, and intellectual property issues can arise.

In a locality consisting group of industries such as agriculture, aquaculture, and energy production, industrial symbiosis could enhance the efficiency and sustainability of each industry through the participation of photosynthetic microbes. Microalgae are cultivated in open ponds or photobioreactors within the biorefinery. They utilize sunlight, carbon dioxide, and nutrients to grow and produce biomass rich in proteins, lipids, and carbohydrates. Emissions of carbon dioxide from nearby industrial sources, such as power plants or manufacturing facilities, can be captured and directed to the microalgae cultivation ponds (Sreeharsha et al. 2023). The microorganisms would use this carbon dioxide as a feedstock for photosynthesis, effectively converting an industrial waste product into valuable biomass. Agricultural runoff or waste streams containing nutrients like nitrogen and phosphorus can be channeled into the biorefinery's microalgae cultivation system. The microorganisms act as nutrient absorbers, cleaning the water and converting the nutrients into biomass. Subsequently, the harvested microalgal biomass can be processed within the biorefinery to extract various valuable components: for instance, lipids can be converted into biodiesel or aviation fuels. Proteins and carbohydrates can be used for animal feed or food supplements. Biomass residues can be converted into biochemicals or bioplastics. Waste heat generated by industries in the region can be utilized to maintain optimal cultivation temperatures for the microorganisms, enhancing their growth rates and productivity. Carbon dioxide, nutrients, and waste heat from one industry become valuable resources for another. By capturing carbon dioxide emissions and reusing nutrients, the biorefinery contributes to greenhouse gas reduction and nutrient management. Collaborative resource exchange enhances the sustainability and economic viability of each participating industry by reducing the waste disposal costs. Industrial symbiosis transforms individual industries into interdependent components of a larger ecosystem, where waste becomes a valuable input and collaboration leads to improved resource efficiency and environmental sustainability. The integration of microbial photosynthesis within a biorefinery framework exemplifies how industries can leverage nature's capabilities to create a more circular, efficient, and resilient industrial landscape.

9.2 Role of Photosynthetic Microbes in Nutrient Recycling

9.2.1 The Carbon Cycles

The surge in the global population is causing severe repercussions on climate change, agricultural land utilization, and waste management, thereby detrimentally impacting food security. Anthropogenic emission of greenhouse gases like carbon dioxide, nitrous oxide, and methane is driving climate change, disrupting Earth's energy equilibrium. As per the IPCC, atmospheric CO_2 concentrations are projected to escalate to around 670–940 μmol mol^{-1} by 2100, nearly doubling the current levels (Stocker et al. 2013). Initially, climate change was thought to affect localized species; however, advanced research, such as Free Air CO_2 Enrichment (FACE) studies, demonstrated its widespread influence on global ecosystem physiology, energy flow, food webs, and biodiversity (Dusenge et al. 2019; Keenan et al. 2016; Leakey et al. 2009; Norby and Zak 2011). Unstable crop yields are merely one facet of climate change's numerous adversities. Moreover, abiotic stresses like drought, salinity, and rising temperatures exacerbate food productivity challenges across the globe. Notably, advancements in plant science have made strides in enhancing food crop resilience to current and future climatic shifts (Ainsworth and Rogers 2007; Jobe et al. 2020; Kant et al. 2012; Palit et al. 2020; Sreeharsha et al. 2015). Developing improved germplasm through conventional and molecular breeding techniques holds promise for addressing dietary needs, malnutrition, and global hunger. Principles of carbon fixation, waste recycling, bioprocessing, and ecological exchange of gases and nutrients can combine to establish effective ecosystems with applications in urban farming and climate mitigation. This demands systematic integration of existing technologies within a sustainable, closed-loop framework, achieving the ideals of the circular bioeconomy. Such ecosystems hold significant potential, spanning from urban farming to comprehensive climate change mitigation. Achieving these goals necessitates the systematic convergence of existing technologies, orchestrated within a sustainable closed-loop framework—a harmonious symphony culminating in the realization of the circular bioeconomy. By integrating microalgae into the intricate web of a bioregenerative urban ecosystem, we can address waste management, food security, and climate mitigation in a synchronized dance of ecological harmony.

9.2.2 Nutrient Cycle

Maintaining consistent growth and productivity requires an ample supply of nutrients and water. Nitrogen, in particular, plays a pivotal role in sustaining photosynthesis, as a substantial portion (about 25–40%) of leaf nitrogen is essential for critical components like RUBISCO and other photosynthetic proteins. As the plant's carbon fixation rate increases, it must strike a balance between increased carbon content and optimal nitrogen levels to ensure steady protein biosynthesis (Sreeharsha et al. 2015). This phenomenon is evident in both C3 and C4 crop plants grown under heightened carbon dioxide levels, where the imbalance between carbon and nitrogen led to photosynthetic acclimation.

The intricate operation of a Balanced Bioenergy System (BBES) necessitates a complex cycle of nutrient recycling (Fig. 9.1). Soil microbiota play a pivotal role in nitrogen assimilation, orchestrating a sequence of events. Key actors in this process include Bacillus, Clostridium, Proteus, Pseudomonas, and Streptomyces bacteria, which undertake ammonification. Similarly, nitrification is catalyzed by species such as *Nitrosomonas*, *Nitrosococcus*, *Nitrobacter*, and *Nitrococcus*. The process of denitrification is executed by organisms like *Thiobacillus denitrificans*, *Micrococcus denitrificans*, and specific species of Serratia, Pseudomonas, and Achromobacter (Zhang et al. 2020).

Upon perceiving organic nitrogen compounds, plants must convert them into readily available forms for protein synthesis and other metabolic functions. A cyclic pathway known as GS-GOGAT governs ammonium assimilation, featuring two critical enzymes: glutamine synthetase (GS) and glutamate synthase (GOGAT). Within this pathway, GS seizes ammonia, converting glutamate into glutamine. Subsequently, GOGAT synthesizes 2-oxoglutarate while also regenerating glutamate.

Rhizobium, another vital group of microbiotas, engages in nitrogen fixation by forming symbiotic relationships with plant roots, leading to the development of root nodules. Notable examples of nitrogen-fixing rhizobia encompass *Sinorhizobium meliloti*, known to nodulate species like *Melilotus*, *Medicago*, and *Trigonella*. Additionally, rhizobium species like *Rhizobium leguminosarum*, *Rhizobium alamii*, *Rhizobium lentis*, *Rhizobium japonicum*, *Rhizobium trifolii*, *Rhizobium phaseoli*, and *Rhizobium smilacinae* exhibit a broad host range.

Phosphorus recycling involves a sequence of processes, including solubilization, depolymerization, mineralization, and assimilation. Bacteria capable of phosphate solubilization comprise *Pseudomonas* Sp., *Bacillus* Sp. (such as *Bacillus circulans* and *Bacillus mucilaginosus*), *Burkholderia* sp., *Paenibacillus* sp., *Acidithiobacillus ferrooxidans*, *Serratia marcescens*, and *Enterobacter* sp. These microorganisms have the remarkable ability to transform insoluble inorganic phosphorus into soluble forms.

The intricate process of carbon recycling can occur under both anaerobic and aerobic conditions, facilitated by various microbial species. Anaerobic conditions involve fermentation, a process involving bacteria such as *Klebsiella*,

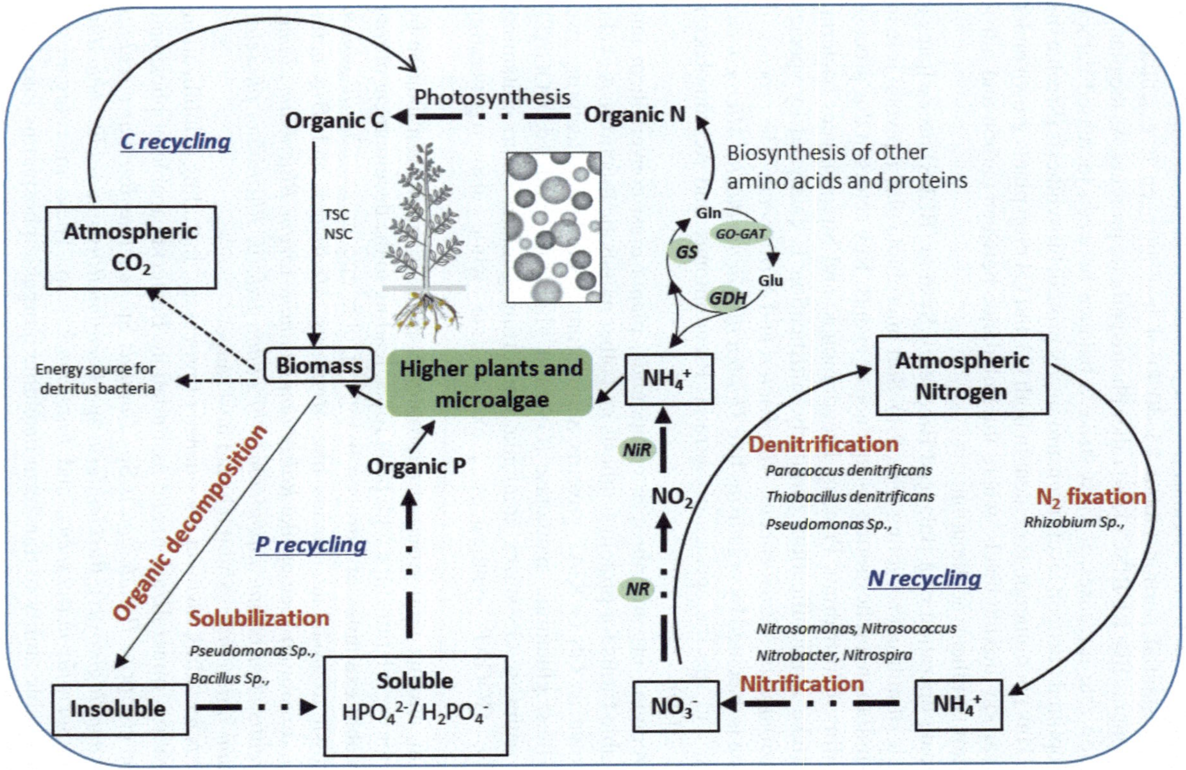

Fig. 9.1 Schematic representation of the symbiotic association of photosynthetic microbes with higher plants and aspects of biome to perform biogeochemical cycles. (Reprinted with permission from Sreeharsha and Venkata Mohan 2021)

Syntrophomonas, *Clostridium butyricum*, and *Bacteroides succinogenes*. Additionally, in anaerobic settings, CO_2 can be converted into methane with the aid of methanogenic bacteria. Conversely, methanotrophs transform methane back into CO_2 under aerobic conditions, creating a cyclic process where carbon is fixed by these bacteria and then recycled. Moreover, bacteria like purple sulfur bacteria and green sulfur bacteria contribute to carbon degradation by breaking down hydrogen sulfide and iron compounds. A prime example includes *Thiobacillus ferrooxidans*.

In summary, the dynamic interplay of microbial interactions is crucial in maintaining nutrient cycles and sustaining the intricate balance of ecosystems. These microorganisms, acting as the architects of nutrient recycling and carbon management, exemplify the intricate web of life's interconnectedness.

9.3 Conclusion

Industrial symbiosis represents a paradigm shift in how industries and economies operate, moving away from linear, wasteful models toward circular and sustainable ones. By fostering collaboration, resource efficiency, and environmental responsibility, industrial symbiosis is playing a pivotal role in shaping a more sustainable and resilient future for industries and communities alike.

References

Ainsworth EA, Rogers A (2007) The response of photosynthesis and stomatal conductance to rising [CO_2]: mechanisms and environmental interactions. Plant Cell Environ 30:258–270

Alberti M (2008) Advances in urban ecology: integrating humans and ecological processes in urban ecosystems, vol 574.5268 A4. Springer, New York

Bamsey M et al (2009) Canadian advanced life support capacities and future directions. Adv Space Res 44:151–161

Chunxiao X, Hong L (2008) Crop candidates for the bioregenerative life support systems in China. Acta Astronaut 63:1076–1080

Dijst M et al (2018) Exploring urban metabolism—towards an interdisciplinary perspective. Resour Conserv Recycl 132:190–203

Dusenge ME et al (2019) Plant carbon metabolism and climate change: elevated CO_2 and temperature impacts on photosynthesis, photorespiration and respiration. New Phytol 221:32–49

Gandy M (2004) Rethinking urban metabolism: water, space and the modern city. City 8:363–379

Jobe TO et al (2020) Ensuring nutritious food under elevated CO_2 conditions: a case for improved C4 crops. Front Plant Sci 11:1267

Kant S et al (2012) Improving yield potential in crops under elevated CO_2: integrating the photosynthetic and nitrogen utilization efficiencies. Front Plant Sci 3:162

Keenan TF et al (2016) Recent pause in the growth rate of atmospheric CO_2 due to enhanced terrestrial carbon uptake. Nat Commun 7:1–10

Leakey AD et al (2009) Elevated CO_2 effects on plant carbon, nitrogen, and water relations: six important lessons from FACE. J Exp Bot 60:2859–2876

Mohan SV et al (2016) Waste biorefinery: a new paradigm for a sustainable bioelectro economy. Trends Biotechnol 34:852–855
Mohan SV et al (2019) Can circular bioeconomy be fueled by waste biorefineries—a closer look. Bioresour Technol Rep 7:100277
Mohan SV et al (2020) Urban biocycles-closing metabolic loops for resilient and regenerative ecosystem: a perspective. Bioresour Technol 306:123098
Norby RJ, Zak DR (2011) Ecological lessons from free-air CO_2 enrichment (FACE) experiments. Annu Rev Ecol Evol Syst 42:181–203
Palit P et al (2020) An integrated research framework combining genomics, systems biology, physiology, modelling and breeding for legume improvement in response to elevated CO_2 under climate change scenario. Curr Plant Biol 22:100149
Restrepo JDC, Morales-Pinzón T (2018) Urban metabolism and sustainability: precedents, genesis and research perspectives. Resour Conserv Recycl 131:216–224
Sreeharsha RV et al (2015) Delayed flowering is associated with lack of photosynthetic acclimation in Pigeon pea (Cajanus cajan L.) grown under elevated CO2. Plant Sci 231:82–93
Sreeharsha RV, Venkata Mohan S (2021) Symbiotic integration of bioprocesses to design a self-sustainable life supporting ecosystem in a circular economy framework. Bioresour Technol 326:124712
Sreeharsha RV et al (2023) Orienting biodiesel production towards sustainability and circularity by tailoring the feedstocks and processes. J Clean Prod:137526
Stocker TF et al. (2013) Climate Change 2013. The Physical Science Basis. Working Group I Contribution to the Fifth Assessment Report of the Intergovernmental Panel on Climate Change-Abstract for decision-makers; Changements climatiques 2013. Les elements scientifiques. Contribution du groupe de travail I au cinquieme rapport d'evaluation du groupe d'experts intergouvernemental sur l'evolution du CLIMAT-Resume a l'intention des decideurs
Zhang Z, Zhang Y, Chen Y (2020) Recent advances in partial denitrification in biological nitrogen removal: from enrichment to application. Bioresour Technol 298:122444

Chapter 10
Artificial Photosynthesis: Nexus of Photon Energy and Bioreactors

Abstract The rising energy demands of the rapidly growing population are a major challenge, involving science and geopolitics. Fossil fuels are diminishing at a faster pace, leading to rising greenhouse gases and temperature. Artificial photosynthesis can contribute to reduce greenhouse gas emissions and dependence on fossil fuels by harnessing solar energy. The key to achieving solar-driven water splitting is to create a light-absorbing material that captures a solar photon and generates a wireless current harnessed by catalysts. The artificial Mn_4CaO_4-complex, which contains a Mn_3CaO_4 cubane attached with a satellite Mn ion, has been found to mimic the atomic structures of the OEC. The oxidation states of the four manganese ions of the Mn_4CaO_4-complex are similar to the S1 state, and the redox transitions suggest that the artificial Mn_4CaO_4-cluster could serve as a catalyst for the water-splitting reaction. The size of the light-harvesting antenna has evolved greatly among photosynthetic organisms, and the size is inversely proportional to light intensity. The CpSRP protein plays a role in reducing antenna size, improving photosynthetic efficiency and growth rates under intense light conditions. Another advantage of photolytic systems is to produce biohydrogen through direct photolysis, involving light reactions of photosynthesis. Biohydrogen can be produced from microalgae through indirect photolysis under sulfur deprivation, which causes an imbalance in photosynthesis and respiration. Strategies to improve biohydrogen production include downregulating photosystem complex proteins and upregulating ATP synthase subunits to divert excess hydrogen ions towards the stroma. However, photodamage to photosystem complexes needs to be addressed through developing an efficient photobioreactor. This chapter gives a basic understanding of the principles and techniques of harnessing of solar energy through artificial antennae for value addition.

Keywords Artificial photosynthesis · Biohydrogen · LHC · Oxygen-evolving complex · Engineered photolysis

R. V. Sreeharsha, S. Venkata Mohan, *Microbial Photosynthesis*,
https://doi.org/10.1007/978-981-97-1253-3_10

10.1 Introduction

The rising energy demands of the rapidly growing population is the major challenge that the current world is facing and it intertwines science and geopolitics. Since centuries, the anthropogenic energy needs are met majorly by fossil fuels nevertheless, they are diminishing at a faster pace and they no longer support the growing needs of the global community. Furthermore, combustion of fossil fuels for energy needs are causing ancillary problems including rising greenhouse gases and temperature. United Nations Framework Convention on Climate Change set the objectives to reduce the greenhouse gas emissions to limit the rise in global temperature not more than 2 °C. However, it raises several economic, environmental, and scientific challenges. A variety of renewable energy sources including wind, biomass, geothermal, hydro, etc. are emerging as promising resources with their own merits and demerits. However, utilizing the energy from the sun is the utmost solution wherein photosynthesis process is crucial in harnessing solar energy.

Photosynthesis evolved billions of years ago and is an indispensable metabolic pathway for producing food and fuels using natural resources. Photosynthesis process in a nutshell, solar energy is converted into chemical energy, and during the process, the leaf splits water molecule to produce molecular oxygen and hydrogen. Oxygen-evolving complex (OEC) located in the vicinity of photosystem II (PSII) plays a major role in generating these segregated protons and electrons. These electrons are then captured by ferredoxin of photosystem I and used to produce hydrogen in the form of NADPH by ferredoxin NADP reductase. Thusly generated reducing equivalents are utilized in later steps to fix the atmospheric CO_2 and produce organic carbon compounds (Sreeharsha et al. 2015). Nature has spatially separated the water splitting and CO_2 reduction in photosynthetic organisms, and such biological engineering aids in utilizing the energy equivalents produced from water splitting to reduce CO_2 and to produce H_2. This kind of structural compartmentalization provides the basis for synthesizing the components to perform artificial photosynthesis which is one of the possible solutions to global warming.

Artificial photosynthesis mimics the light and dark reactions of natural photosynthesis and performs photon collection, and CO_2 captures thus ultimately producing energy (electricity and hydrogen) and biopolymers (including food) with high efficiency. This type of bio analog system exhibits several advantages including non-polluting electricity generation, photohydrogen and alternative carbon products generation, and carbon dioxide sequestration. In the process, these artificial photosynthetic technologies significantly reduce the greenhouse gas emissions and our dependence on fossil fuels. In principle, the artificial photosynthesis system has to mimic nature's method of water splitting, electron transfer, and chemical/energy production in an engineered platform. Consequently, the key to simulate photosynthesis lies largely in the ability to achieve solar-driven water splitting by a direct method. To realize this, there must be a light-absorbing material that captures a solar photon to generate a wireless current that is harnessed by catalysts, which fulfills the function of water-splitting reaction and electron generation under benign conditions. One of the major challenges for scientists is to explore for the compounds which effectively equate the oxygen and hydrogen-evolving complex and match the properties of water splitting of the natural photosynthetic reaction. The other challenge is to integrate these compounds with a light-harvesting semiconductor which ultimately creates an artificial leaf and permits direct solar-to-fuels conversion using one sun illumination (100 mW/cm^2). Producing H_2 gas and other carbon-based fuels through artificial photosynthesis poses several advantages in terms of storage and energy extraction as the cost of storing energy as hydrogen is cheaper than storing it as electricity. This chapter discuss various strategies to harness solar energy through designing artificial light-harvesting systems and catalysts to mediate electron transport. Also, production of biohydrogen and various strategies to improve biohydrogen production from microalgae are discussed.

10.2 Advancements in Designing Artificial Photosynthetic Components

10.2.1 Catalysts That Mimic OEC

The mechanism involved in the splitting of water molecule by OEC is still far from understanding because of the complex structural properties of the proteins and uncertainty of the intermediate states of the OEC during its turnover. So, the crucial aspect of artificial photosynthesis is to engineer a catalyst which mimics the OEC and catalyze water-splitting reaction. Crystallographic studies of OEC of various microbial and plant species help in understanding its detailed intricacies of structural and functional aspects which in turn aided chemists to synthesize the OEC (Umena et al. 2011). For several years, it has been a challenging task to incorporate Ca^{2+} into the Mn^{4-} cluster through μ-oxo bridges, to match the asymmetric Mn_4Ca-cluster of OEC because the affinity of Ca^{2+} to μ-oxo is significantly weaker than that of Mn ion (Zhang et al. 2015). Another challenge is to match the pyridine-ligands to that of natural OEC ligands which are majorly carboxylate groups and water molecules (Dismukes et al. 2009; Karkas et al. 2014; Mullins and Pecoraro 2008; Najafpour et al. 2016a; Tsui and Agapie 2013). Several groups have reported successful synthesis of Mn clusters with diverse peripheral ligands wherein the tetra-manganese complexes containing Mn_4O_4-cubane forms the basis for all successive developments (Chakov et al. 2016; Kanady et al. 2011). Artificial Mn_3CaO_4-cluster using a multipyridyl-alkoxide ligand, and the $Mn_3Ca_2O_4$-complex with neutral pivalic acid ligands that are synthesized by attaching additional Ca^{2+} to the Mn_3CaO_4 cubane are the popular examples that closely mimic the peripheral carboxylate ligands of the OEC in PSII (Kanady et al. 2011). While the artificial Mn_3CaAgO_4-complex is another example of artificial system which showed considerable potential in water-splitting reaction. In general, artificial complexes contain Mn ions in oxidation state IV, with varying bond lengths and ion distances. Typically, Mn-O and Ca-O bond lengths are in the range of 1.8–1.9 and 2.4–2.7 Å, respectively, while Mn-Mn and Mn-Ca distances are in the range of 2.7–2.8 and 3.2–3.5 Å, respectively. With these structural features, synthesizing the entire Mn_4Ca cluster with ligands having similar properties as seen in the natural OEC remains a challenge.

The core structure of the artificial Mn_4CaO_4-complex contains a Mn_3CaO_4 cubane attached with a satellite Mn ion and surrounded by pyridine and pivalic acid ligands. Pertinently, absence of water as solvent in the crystal of artificial Mn_4CaO_4-cluster limits the effect of the X-ray radiation reduction. Based on the Bond Valence Sum (BVS) calculations and cyclic voltammogram (CV) measurements of the artificial Mn_4CaO_4-complex, it is clear that the oxidation states of the four Mn ions are III, III, IV, IV which are similar to that of S1 state. Also, artificial Mn_4CaO_4-complex and natural OEC are similar in terms of redox transition and redox potential thus serving as a catalyst for the water-splitting reaction (Dau and Zaharieva 2009; Boussac 2019; Pantazis 2021). The artificial Mn_4CaO_4-complexes can also catalyze oxygen-evolving reaction efficiently by using $(CH_3)_3COOH$ (tert-butyl hydroperoxide) as an oxidant in acetonitrile with the presence of 4.5% water. However, these artificial complexes are very sensitive to experimental conditions. Quantification of the catalytic reaction has been difficult due to the rapid degradation of the catalyst in the presence of water in solution, making it difficult to quantify the catalytic reaction. The calcium in the artificial complex has been found to be easily dissociated, leading to the formation of multi-manganese complexes. The stability of the artificial Mn_4CaO_4-cluster can be enhanced using acetonitrile and N, N-dimethylformamide (DMF), as exchangeable solvent on the calcium. Remarkably, the replacement of one or two ligands on the calcium by organic solvent molecules does not modify the core structure and oxidation valences of the four manganese ions, as well as main peripheral environment ligands. More importantly, these new Mn_4CaO_4-clusters become more stable in the polar solvent, which may provide great opportunity to investigate the catalytic activity of the artificial Mn_4CaO_4-cluster in the future. Nevertheless, artificial Mn_4CaO_4-clusters are thermodynamically stable and provide distinct structural insight into understanding the principle of the OEC (Barber 2016, 2017; Kawashima et al. 2018).

10.2.2 Modulating Antenna Size

The size of the light-harvesting antenna has been evolved greatly among photosynthetic organisms from lower microbes to higher plants. The size of the antenna has a great impact on the amount of light that is harvested and in turn decides the photooxidative damage which is caused due to pigment bleaching due to excessive light. Hence, the antenna size is a kind of photoprotective mechanisms that the organisms have developed. The synthesis of light-harvesting complex is a regulated process that is determined by various genes including

NAB1, CAO, and other light-harvesting proteins (Dall'Osto et al. 2015). In general, the antenna size is inversely proportional to the light intensity, and it is species specific. The antenna size can be seen as an adaptive mechanism to protect against photooxidation rather than the photosynthetic efficiency and pigment composition. Usually, microalgae are evolved to have bigger antenna sizes compared to land plants due to their habitats of low-light conditions (Perrine et al. 2012). When the overall cultivation efficiency is concerned, the bigger antenna size is proportionate to excess loss of energy and hence manipulation of antenna size is essential to enhance the culture efficiency. The initial attempts to reduce the antenna size was mostly focused on creating *Chlamydomonas* mutants through random mutagenesis, and this led to the identification of the role of chloroplast signal recognition particle (CpSRP) in reducing the antenna size. CpSRP proteins are involved in folding and translocation of LHC proteins to the lipid bilayer of thylakoid membrane and any mutation in CpSRP-encoding gene often results in antenna size reduction (Kirst et al. 2014). The CpSRP mutants in *Chlamydomonas* showed reduction in PSII antenna size and increase in the chlorophyll *a/b* ratio, which showed enhanced photosynthesis and biomass accumulation under intense light conditions. Whereas *Chlorella* CpSRP43 mutants showed decrease in photosynthetic efficiency and growth rates under low-light conditions (Shin et al. 2016).

10.3 Biohydrogen Production Through Photolysis

Biohydrogen production through biological process is cost-intensive and more sustainable when compared to the other physical and chemical techniques (Islam et al. 2021; Raghulchandrana et al. 2020). The biological process of hydrogen production includes dark fermentation, photo-fermentation, and direct photolysis. Dark fermentation utilizes organic derivatives from sugar hydrolysis and produce biohydrogen in the absence of light using certain fermentative microorganisms including *Clostridium beijerinckii*, *Clostridium butyricum*, *Enterobacter aerogenes*, and *Enterobacter asburiae*. DF is regarded as a very feasible technique of H_2 production with greater efficiencies under optimized physicochemical parameters like temperature, pH, hydraulic retention time, inoculum amount, and size.

In DF, H_2 production is mediated by two key enzymes including hydrogenase and nitrogenase (Mishra et al. 2019; Najafpour et al. 2016b). Hydrogenases which catalyze the inter-conversion of H_2 to protons and electrons are categorized into three classes, namely [NiFe]-hydrogenase, [FeFe]-hydrogenase, and [Fe]-hydrogenase. These enzymes participate actively in the acetate and butyrate pathways of DF which gives theoretical yield of 4 and 2 mol H_2 from 1 mol glucose, respectively (Islam et al. 2021; Sarangi and Nanda 2020; Sekoai and Daramola 2018). In the first stage of DF, NADH is produced via oxidation of substrate into pyruvate and may be utilized by microorganisms having NADH:

ferredoxin oxidoreductase (NFOR) to produce reduced ferredoxin (Preethi 2023). While in the second stage, pyruvate can be converted into acetyl-CoA and formate by pyruvate formate lyase (PFL) or acetyl-CoA and reduced ferredoxin via pyruvate-ferredoxin oxidoreductase (PFOR) to produce H_2 (Sağır and Hallenbeck 2019; Zhu and Xu 2015). Generation of acids and acid metabolites including ethanol, acetic acid, fumaric, lactic, butyric acid, and propionic acid lowers the yield of H_2 and is considered as one of the biggest challenges for DF (Sarangi and Nanda 2020). Also, formation of large amount of volatile fatty acids (VFA) makes the fermentation medium more acidic and hamper the H_2 production (Bhatia et al. 2021; Mishra et al. 2019; Osman et al. 2023; Slezak et al. 2021). However, integrating the dark fermentation with photo-fermentation makes better utilization of these short fatty acids as substrates. There are good efforts to develop a cumulative bioprocess by integrating these both technologies to produce high biohydrogen (Hajizadeh et al. 2021; Mishra et al. 2019; Soares et al. 2020; Yin and Wang 2019).

10.3.1 Photo-Fermentation

Photo-fermentation is majorly carried out by photosynthetic purple non-sulfur (PNS) bacteria in which volatile fatty acids are converted into H_2 in the presence of nitrogenase enzyme in the presence of light source (Reungsang et al. 2018; Pandey et al. 2023). Photosynthetic bacterial species including *Rhodobacter sphaeroides*, *Rhodobacter capsulatus*, *Rhodovulum sulfidophilum* W-1S, and *Rhodopseudomonas palustris*, *Rhodobacter* are well known for H_2 production via photo-fermentation using VFAs as substrate (Mishra et al. 2019). Whereas acetic acid, butyric acid, propionic acid, malic acid, and lactic acid are VFAs that act as potential substrates for photo-fermentative H_2 production (Assawamongkholsiri et al. 2019). Other than VFAs, sugars such as glucose and sucrose also serve as substrates for photo-fermentative H_2 production (Rai et al. 2014). Additionally, effluent of the DF can be directly used for the photo-fermentation in the sequential process without any pH change. Along with substrate and microbes, optimization of physicochemical parameters is also crucial for the high yield of H_2 (Table 10.1) (Soto et al. 2019).

Table 10.1 Optimum parameters for effective production of H_2 through photo-fermentation

Parameter	Range
Temperature	30–40 °C
pH	5.0–7.0
Pressure	Atmospheric
Light (nm)	522, 805, 850

10.3.2 Direct Photolysis

Light reactions of photosynthesis lead to generation of electrons during water-splitting process which are transferred to [Fe]-hydrogenase through PSI electron carriers, leading to hydrogen production. The protein–pigment complexes of PSII located in the thylakoid membrane play a crucial role in the generation of biohydrogen. PSII-specific water-plastoquinone oxidoreductase and PSI-specific ferredoxin oxidoreductase of microalgae and cyanobacteria are known to act as rate-limiting enzymes in this process. When the photosystem complexes absorb light energy, the generated electrons form hydrogen ions and molecular oxygen (Chap. 2). These hydrogen ions generate a proton gradient which is used by ATP synthase to generate ATP and simultaneously release hydrogen gas through the action of [Fe]-hydrogenase (Fig. 10.1). The ability of microorganisms to produce hydrogen gas capitalizes on their light harvesting and water-splitting efficiency which in turn depends on the structural chemistry of photosystem complexes. When the photosystems are exposed to excess light, the resultant hydrogen ions are diverted towards hydrogen gas nevertheless, excess light causes photoinhibition by damaging photosystem complexes.

Production of hydrogen gas through direct photolysis offers several advantages when compared to the other two methods as the process relies majorly on naturally available water and sunlight making the process more sustainable. However, simultaneous production of oxygen and hydrogen gases is the major challenge as oxygen is the strong inhibitor of the hydrogenase enzyme. Another challenge is to design suitable bioreactors to cultivate microalgae in large scale that utilize solar energy effectively to produce sustainable biohydrogen.

Biohydrogen can be produced from microalgae through indirect photolysis under sulfur deprivation which causes reversible inhibition in the activity of oxygenic photosynthesis by impeding protein biosynthesis and D1 turnover

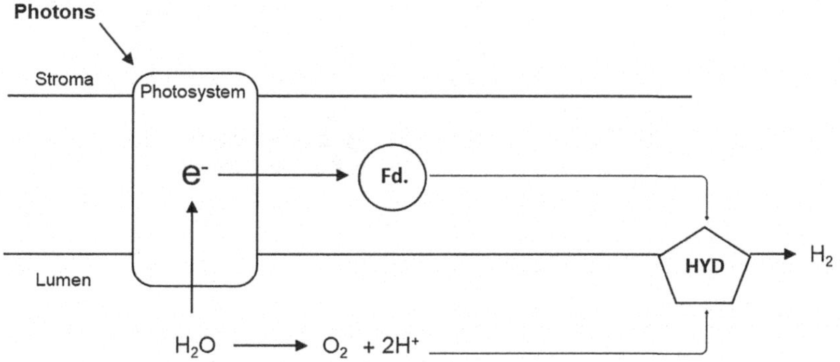

Fig. 10.1 Schematic representation of biohydrogen production by direct photolysis method

which is essential for functioning of PSII. This imposes a situation where the net photosynthetic capacity and O_2 release becomes lower than the rate of O_2 consumption in respiration. This imbalance in photosynthesis and respiration caused by sulfur deprivation leads to the generation of anaerobic conditions in the growth medium.

10.4 Strategies to Improve Biohydrogen Production

Photosystem complex proteins are the ideal targets to improve biohydrogen production. When exposed to high light, Psbs protein dissociates from the LHC1 and initiates heat dissipation in the form of NPQ. The excess electrons generated due to the high light exposure are released into the stroma which subsequently unutilized for downstream applications. Downregulating Psbs with simultaneous upregulation of ATP synthase subunits could be an effective strategy to divert the excess hydrogen ions towards stroma via ATP synthase (Fig. 10.2). These electrons will be subsequently transferred to ferredoxin and hydrogenase enzyme to produce enhanced biohydrogen. However, the photodamage to the photosystem complexes needs to be addressed through developing an efficient photobioreactor which allows the microalgae to expose intermittent high light conditions in a controlled manner.

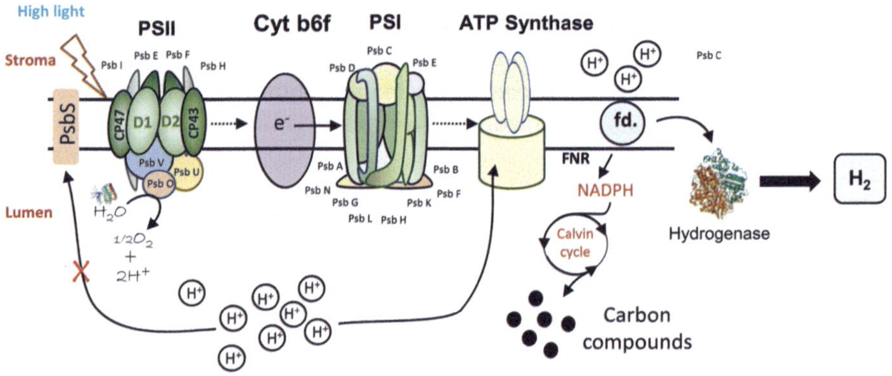

Fig. 10.2 Schematic representation of engineering strategy to improve biohydrogen production by direct photolysis method in microalgae or cyanobacteria

10.5 Conclusions

Harnessing solar energy through artificial systems is a promising technology for renewable energy production. The primary events that have to be fulfilled by an artificial photosynthesis frameup include light absorption, electron transport, and ultimate production of biohydrogen or electricity. The developments in the material science and chemical engineering enable the synthesis of catalysts and nanomaterials which mimic OEC and light-harvesting complexes with similar structural and functional properties. Also, microalgae and cyanobacteria photosystems are the source of engineering for efficient biohydrogen production. Identifying candidate proteins in photosystem to minimize NPQ and enhanced biohydrogen production could be ideal strategy to meet the demands of the biohydrogen which is the fuel for the future.

References

Assawamongkholsiri T et al (2019) Photo-hydrogen and lipid production from lactate, acetate, butyrate, and sugar manufacturing wastewater with an alternative nitrogen source by Rhodobacter sp. KKU-PS1. PeerJ 7:e6653

Barber J (2016) Photosystem II: the water splitting enzyme of photosynthesis and the origin of oxygen in our atmosphere. Q Rev Biophys 49:e14

Barber J (2017) A mechanism for water splitting and oxygen production in photosynthesis. Nat Plants 3:1–5

Bhatia SK et al (2021) Renewable biohydrogen production from lignocellulosic biomass using fermentation and integration of systems with other energy generation technologies. Sci Total Environ 765:144429

Boussac A (2019) Temperature dependence of the high-spin S2 to S3 transition in photosystem II: mechanistic consequences. Biochim Biophys Acta Bioenerg 1860:508–518

Chakov NE et al (2016) Unusual MnIII/IV4 cubane and MnIII16M4 (M= Ca, Sr) looplike clusters from the use of dimethylarsinic acid. Inorg Chem 55:8468–8477

Dall'Osto L et al (2015) Biogenesis of light harvesting proteins. Biochim Biophys Acta Bioenerg 1847:861–871

Dau H, Zaharieva I (2009) Principles, efficiency, and blueprint character of solar-energy conversion in photosynthetic water oxidation. Acc Chem Res 42:1861–1870

Dismukes GC et al (2009) Development of bioinspired Mn_4O_4—cubane water oxidation catalysts: lessons from photosynthesis. Acc Chem Res 42:1935–1943

Hajizadeh A et al (2021) Biohydrogen production through mixed culture dark anaerobic fermentation of industrial waste. In: Integrated natural resources management. Springer, Cham, pp 323–369

Islam AK et al (2021) Bio-hydrogen production from wastewater: a comparative study of low energy intensive production processes. Clean Technol 3:156–182

Kanady JS et al (2011) A synthetic model of the Mn_3Ca subsite of the oxygen-evolving complex in photosystem II. Science 333:733–736

Karkas MD et al (2014) Artificial photosynthesis: from nanosecond electron transfer to catalytic water oxidation. Acc Chem Res 47:100–111

Kawashima K et al (2018) O2 evolution and recovery of the water-oxidizing enzyme. Nat Commun 9:1247

Kirst H et al (2014) Maximizing photosynthetic efficiency and culture productivity in cyanobacteria upon minimizing the phycobilisome light-harvesting antenna size. Biochim Biophys Acta Bioenerg 1837:1653–1664

Mishra P et al (2019) Outlook of fermentative hydrogen production techniques: an overview of dark, photo and integrated dark-photo fermentative approach to biomass. Energ Strat Rev 24:27–37

Mullins CS, Pecoraro VL (2008) Reflections on small molecule manganese models that seek to mimic photosynthetic water oxidation chemistry. Coord Chem Rev 252:416–443

Najafpour MM et al (2016a) Engineered polypeptide around nano-sized manganese–calcium oxide as an artificial water-oxidizing enzyme mimicking natural photosynthesis: toward artificial enzymes with highly active site densities. Int J Hydrog Energy 41:17826–17836

Najafpour MM et al (2016b) Manganese compounds as water-oxidizing catalysts: from the natural water-oxidizing complex to nanosized manganese oxide structures. Chem Rev 116:2886–2936

Osman AI et al (2023) Critical challenges in biohydrogen production processes from the organic feedstocks. Biomass Conv Bioref 13:8383–8401

Pandey AK et al (2023) Machine learning in fermentative biohydrogen production: advantages, challenges, and applications. Bioresour Technol 370:128502

Pantazis DA (2021) Natural and artificial photosynthesis. In: Solar-to-chemical conversion: photocatalytic and photoelectrochemcial processes. Wiley, Hoboken, pp 41–76

Perrine Z et al (2012) Optimization of photosynthetic light energy utilization by microalgae. Algal Res 1:134–142

Preethi V (2023) Solar hydrogen production in India. Environ Dev Sustain 25:2105–2135

Raghulchandrana R et al (2020) Pretreatment studies of biohydrogen production from agro-industrial waste. Preprints. https://doi.org/10.20944/preprints202011.0320.v1

Rai PK et al (2014) Biohydrogen production from sugarcane bagasse by integrating dark-and photo-fermentation. Bioresour Technol 152:140–146

Reungsang A et al (2018) Hydrogen from photo fermentation. In: Bioreactors for microbial biomass and energy conversion. Springer, Singapore, pp 221–317

Sağır E, Hallenbeck PC (2019) Photofermentative hydrogen production. In: Biohydrogen. Elsevier, Amsterdam, pp 141–157

Sarangi PK, Nanda S (2020) Biohydrogen production through dark fermentation. Chem Eng Technol 43:601–612

Sekoai PT, Daramola MO (2018) Effect of metal ions on dark fermentative biohydrogen production using suspended and immobilized cells of mixed bacteria. Chem Eng Commun 205:1011–1022

Shin W-S, Lee B, Jeong B-r, Chang YK, Kwon J-H (2016) Truncated light-harvesting chlorophyll antenna size in Chlorella vulgaris improves biomass productivity. J Appl Phycol 28:3193–3202

Slezak R et al (2021) Influence of initial pH on the production of volatile fatty acids and hydrogen during dark fermentation of kitchen waste. Environ Technol 42:4269–4278

Soares JF et al (2020) Dark fermentative biohydrogen production from lignocellulosic biomass: technological challenges and future prospects. Renew Sust Energ Rev 117:109484

Soto LR et al (2019) Hydrogen and polyhydroxybutyrate production from wheat straw hydrolysate using Caldicellulosiruptor species and *Ralstonia eutropha* in a coupled process. Bioresour Technol 272:259–266

Sreeharsha RV et al (2015) Delayed flowering is associated with lack of photosynthetic acclimation in Pigeon pea (Cajanus cajan L.) grown under elevated CO2. Plant Sci 231:82–93

Tsui EY, Agapie T (2013) Reduction potentials of heterometallic manganese–oxido cubane complexes modulated by redox-inactive metals. Proc Natl Acad Sci 110:10084–10088

Umena Y, Kawakami K, Shen J-R, Kamiya N (2011) Crystal structure of oxygen-evolving photosystem II at a resolution of 1.9 Å. Nature 473(7345):55–60

Yin Y, Wang J (2019) Mechanisms of enhanced biohydrogen production from macroalgae by ferrous ion: insights into correlations of microbes and metabolites. Bioresour Technol 291:121808

Zhang C et al (2015) A synthetic Mn4Ca-cluster mimicking the oxygen-evolving center of photosynthesis. Science 348:690–693

Zhu Q-L, Xu Q (2015) Liquid organic and inorganic chemical hydrides for high-capacity hydrogen storage. Energy Environ Sci 8:478–512

The manufacturer's authorised representative in the EU is Springer Nature Customer Service Centre GmbH, Europaplatz 3, 69115 Heidelberg, Germany. If you have any concerns regarding our products, please contact ProductSafety@springernature.com

Printed and bound by CPI Group (UK) Ltd, Croydon, CR0 4YY
01/10/2026
02225862-0002